The Fathom Guide
for

The Practice of Statistics

Third Edition

TI83/84/89 Graphing Calculator Enhanced

TIM ERICKSON

W. H. Freeman and Company
New York

Design and layout by **eeps** media, Oakland, California

ISBN-13: 978-0-7167-8312-1
ISBN-10: 0-7167-8312-6

Second printing

W. H. Freeman and Company
41 Madison Avenue
New York, NY 10010
Houndmills, Basingstoke RG21 6XS, England

www.whfreeman.com

Contents

Introduction: How To Use This Guide

This guide is a supplement to *The Practice of Statistics, Third Edition.* It's all about using *Fathom™ Dynamic Data Software* as you use that book to learn statistics. This guide does not stand alone. You also need a copy of *The Practice of Statistics,* and you need a copy of Fathom.

From now on, we'll refer to *The Practice of Statistics, Third Edition* simply as "the textbook." It's divided into a preliminary chapter and 15 regular, numbered chapters. There is a chapter in this guide corresponding to each chapter in the textbook; within each chapter are numbered examples and exercises that correspond as well.

Throughout this guide, you'll learn to use Fathom to accomplish various statistical tasks and to do the examples and exercises. You'll start out with simple things like making graphs and work up to amazing constructions such as simulations that help you understand statistical tests. Fathom will help you understand some of the toughest parts of statistics, but it can't understand for you: Even though Fathom is easy to learn and use, you'll need to expend some effort to get good at it.

It will be worth it. And you're probably used to learning new programs. If you don't read manuals, don't worry: You can learn a lot of it by just trying stuff out. If you add talking to your classmates and sharing what you discover, you can do pretty well. But don't neglect online help, and don't neglect this guide. As with any sophisticated piece of software, there are usually two or three different ways to accomplish something; this guide will show you some of the most efficient ways to do what you need to do.

We will give you some initial advice about Fathom in a moment, but first a few words about this guide.

Book Conventions

There are several kinds of lists in this guide:

❖ This kind, with the diamond mark, is a *step*. It's something that you do on the computer, such as choose **Paste** from the **Edit** menu.

> ✐ If you see a pencil instead of a diamond, it is a *question* for you to answer or a *task* to do that does *not* appear in the textbook. For example, discuss with your classmates: Why was **Paste** set in that type? (Because you would see the word on the computer screen itself. We set filenames and computer text in this **bold, dark gray** type.)

17.2 THIS IS AN EXERCISE FROM THE TEXTBOOK. If there were an Exercise 17.2 in the textbook (there isn't one), this paragraph and what follows would give you advice on how to use Fathom to do the exercise. In general, you are still responsible for doing the *whole* exercise even though the whole exercise will not appear in this guide. You still need the textbook.

❑ This is for a list that's just a list. Sometimes you can think of it as a check box.

From time to time, we will set off hints between horizontal lines, or important principles and digressions inside gray boxes.

Installation

It will be best to install Fathom, although if you have it on a CD you can run it without installation. Use the installation instructions that come with the software. Your life will be simpler if you make a shortcut somewhere so you can access it easily. For Windows users, that might be in your **Start** menu or on your desktop; for Mac users, that might be in your dock.

Windows and Macintosh

While we're at it, let's talk about the Windows/Mac thing. Fathom runs on both Windows and Mac systems. Although it runs the same and looks very similar, there are small differences in the interface.

1. Keyboard shortcuts on Windows usually use the CONTROL or CTRL key. The same shortcut on the Macintosh uses the "command" key, *not* the control key. The command key on the Mac is the one with ⌘ on it. For example, **Paste** is CONTROL-**V** (Windows) or ⌘-**V** (Macintosh).

2. Fathom makes extensive use of context menus, for which you need the right mouse button. Most standard Macintosh mice have one button. To get the right button, hold down CONTROL (not ⌘!) before you click.

   We will not explain this every time! If you use a one-button Mac mouse, you will have to get used to the instruction "right-click on the graph...." Mentally translate it to "hold down CONTROL and click on the graph...."

There are a few other small keyboard things having to do with ALT and OPTION keys, but we'll deal with them when they come up. When in doubt, check **Help**.

Initiative

The next topic is initiative. This guide assumes that you will take it. For example, we will write "check **Help**." We will not always write, "look in the menu bar at the top of your screen (Mac) or at the top of your application window (Windows) and choose **Fathom Help** from the **Help** menu." We assume that you can visually scan the screen, spot the word **Help** and follow your nose to the right command. Of course you will get detailed instructions the first time you do something hard. But they quickly will get less detailed, because we assume that you're learning it and don't need to be told every step. If you don't remember, it's not a big deal: Refer back to where you learned it, ask your friends, or check **Help**.

The Fathom Window and Fathom Objects

Let's look at a Fathom window and see what's in it. You might learn all this by osmosis, but just in case you want a preview, here we go:

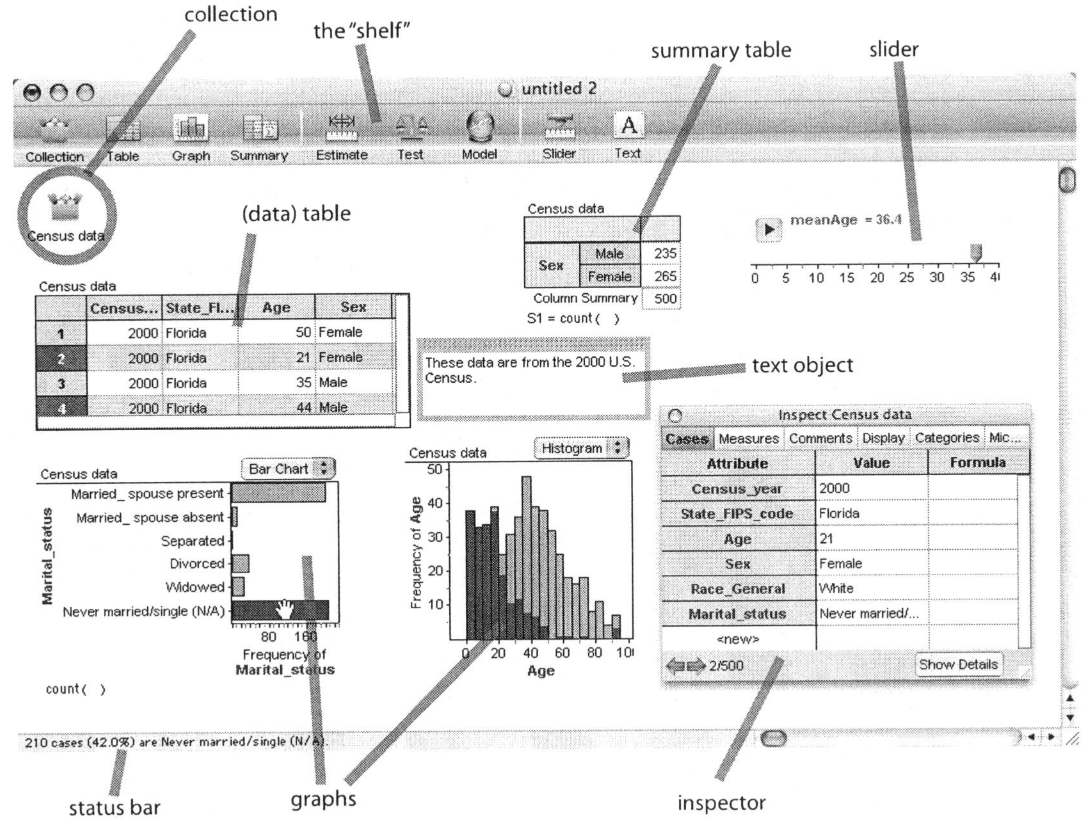

Fathom starts with an empty window. You make new objects by dragging them off the *shelf*, a bar across the top of the window. These objects include graphs, data tables (which are like spreadsheets), summary tables, sliders, and text. You can select these objects (at which point they get a border like the text object above) and resize them. Delete them by pressing DELETE or BACKSPACE. They are not, technically, windows in their own right.

Right-clicking on an object (CONTROL-click on the Mac, remember?) brings up a *context menu*, which has commands appropriate to the object. You'll use context menus a lot. *Double*-clicking a collection, graph, or slider brings up an *inspector*, which is a window that floats over the document. You'll use inspectors a lot, too.

The bottom of the window has a *status bar*. It shows information about whatever the mouse is pointing at. In this case, the mouse (it looks like a hand) is hovering over a bar showing the never-married people. The status bar tells us there are 210 of these.

The *collection* is the most important object. It looks like a box of balls. *It is where the data live.* Every other display is just a *view* of the data. If you delete a table that shows the data, you can make a new one. If you delete the collection, the data are gone.

Can you bring them back? Sure. Fathom supports multiple **Undo** (CONTROL-**Z**, ⌘-**Z**) commands. **Undo** is your friend.

Formulas

About half of your interaction with Fathom will be by dragging stuff and seeing the results. The other half will be writing the formulas that control what's going on. You will do this using Fathom's *formula editor.*

It can be tricky, sometimes, to write the formula the way you want. Some advice: Use parentheses whenever you're not sure. Read the embedded documentation about functions. And when things get really messed up, it's usually not too hard just to start over.

One thing to note: When you're using the formula editor, *you can't do anything else.* Click **OK** to close the formula editor, and then you can go on.

Files

There are many data files on the CD that may come with the textbook, or that you may download. Find the folder of Fathom data. Copy it to your hard disk if you wish. Inside that folder are folders for each chapter. The files you'll need are inside those "chapter" folders. Names start with **EG** for examples (as in "e.g.") and **EX** for exercises. So **EX02_05 Unemployment.ftm** is a file for Exercise 2.5.

Almost Final Words

Fathom will give you unprecedented power to explore data and learn more deeply about statistics. Try to be flexible; there are lots of ways to do things and many discoveries to make. Make more graphs. Select things and see what lights up. Drag sliders. Drag data. It's not called *dynamic* data analysis for nothing. It should be fun.

Acknowledgments

I am grateful to Laura Hanrahan, Brendan Cady, and Craig Bleyer at W. H. Freeman for working with me on this project and getting me everything I needed in a timely fashion. Of course, it would have been completely impossible without the authors of *The Practice of Statistics*, especially Dan Yates and Daren Starnes, who encouraged Freeman to have me do this and then put up with my many e-mailed questions. They also set a high standard for quality and industry, as any user of the gold—wait, now platinum!—binder knows.

On the Fathom side, I sit in continued awe of Bill Finzer, Kirk Swenson, Matt Litwin, Vishakha Parvate, Nick Jackiw, and the rest of the team at KCP Technologies, who make the promise of dynamic data analysis a reality.

Finally, I want to thank Jonah Lipsitt for a conscientious technical review of the early chapters, and of course my family, Meg Holmberg and Anne Erickson, for editing, comments, and putting up with yet another book project.

Tim Erickson
Oakland, California
June 2006.

Preliminary Chapter

The preliminary chapter of *The Practice of Statistics* gives you an overview and introduces some important ideas. This chapter corresponds to that chapter in your textbook. Here, we'll see how some of those statistical ideas play out in Fathom. Along the way, you'll learn the basics of Fathom—how to get some data, display it in a table, and graph it. Subsequent chapters will build on these Fathom skills.

Introduction to Fathom

The preliminary chapter in *The Practice of Statistics* describes how data are organized and what kinds of data you will use. Let's use Fathom to get a data set to play with.

❖ Start Fathom. Usually this will be by double-clicking its icon, which looks like a gold ball. Depending on how your system is set up, it might be in your **Start** menu. Fathom starts with an empty document.

❖ Maximize the size of your window. Fathom uses lots of screen space; give yourself as much as possible.

With Internet

What you do next depends on your situation. If you are connected to the Internet, do the next two steps. If anything doesn't work, skip to the no-Internet alternative.

❖ In the **File** menu, choose **Import**, and then **U. S. Census Data....** A box appears (it's called an *inspector*). (In the future, we'll say "Choose **File>Import>U. S. Census Data....**")

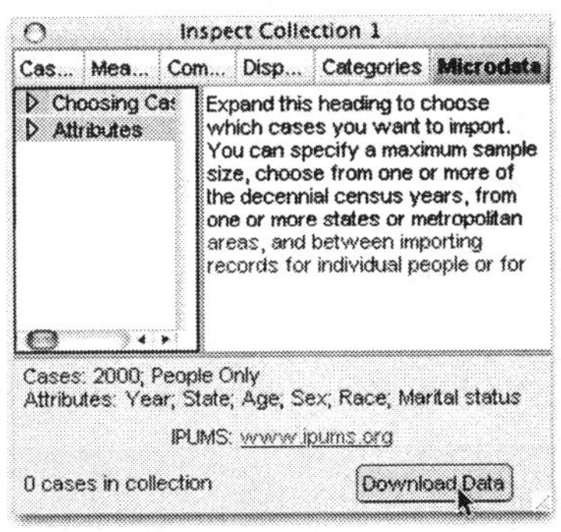

❖ Press the **Download Data** button. The box will tell you that Fathom is contacting a server,[1] polling the server, downloading results, etc., until the box reads **500 cases in collection** at the lower left.

Without Internet

If, on the other hand, you are *not* connected to the Internet (or something went wrong when you tried the steps above):

❖ Open the file **0A Introductory Census Data.ftm**. (We made this file by doing the steps above, so you can continue with the next instructions. You will find it in a folder called **00 Preliminary Chapter**.)

1 The server is part of the IPUMS project at the University of Minnesota. Visit **www.ipums.org** to find out more.

Onward!

What just happened? You got a data set of 500 randomly selected people from the 2000 United States Census. The box of gold balls that appeared, probably called **Collection1**, actually contains the data. Let's look at it in a *table.*

- ❖ Close the inspector (the box with the **Download Data** button).

- ❖ Drag a *table* off the *shelf* and drop it in your document. It will fill with data, as in the left-hand illustration, if the collection (the box of balls) was *selected.* If not, "connect" the collection to the empty table by dragging the *name* of the collection (not its icon) from the collection to the middle of the table, as shown in the right-hand illustration.

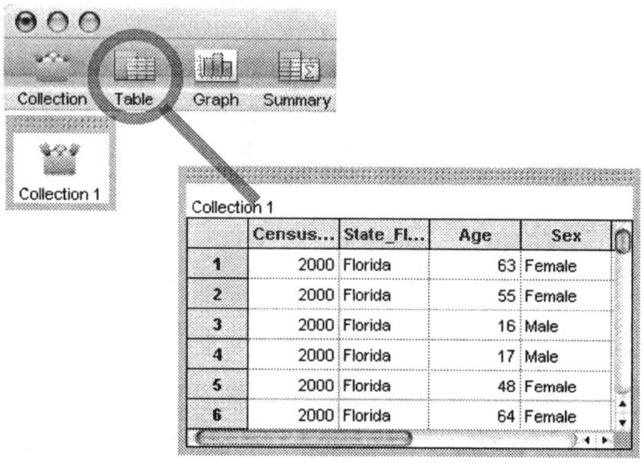

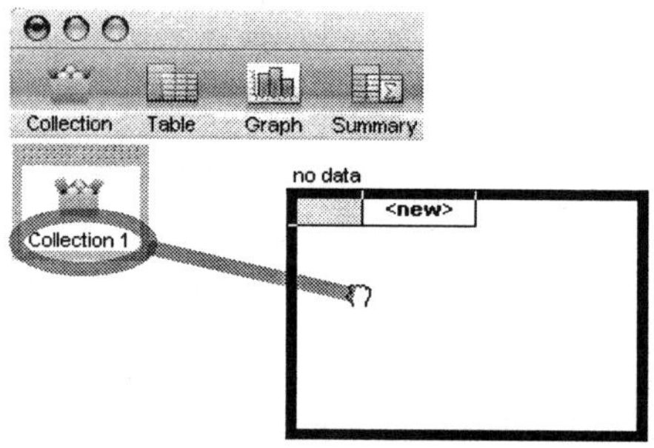

You can move and resize the table in the obvious ways. You can also change the widths of columns. You will probably want to make the table wider, and stretch out the **Marital_status** column so you can read the entries.

- ❖ Scroll around in the table to see what you find. Each row is a person.

- ❖ When you're ready to move on, make the table *short* so you can see all of the column headings but maybe only a row or two of data. If you leave the table huge, you may not have enough space for the next part.

Making graphs in Fathom

Scrolling around in the table is interesting, but it's hard to get a sense of the whole data set because there are so many cases. We need *graphs.*

- ❖ Drag a graph off the shelf, and drop it in an empty part of your window (i.e., not on top of the table). The graph should be empty to begin with.

- ❖ Drag the column heading **Age** from the table to the horizontal axis of the graph (where it says, "**drop an attribute here**"). You can think of this as labeling the axis. Fathom makes a *dot plot* of the ages. There is one dot for each of the 500 people in the collection.

- ❖ Drag another graph off the shelf, and drop it into an empty part of your window.

❖ This time, drag the heading for **Marital_status** and drop it on the vertical axis of the new graph. Fathom makes a *bar chart* of the marital status of the 500 people. Your document should look something like this:

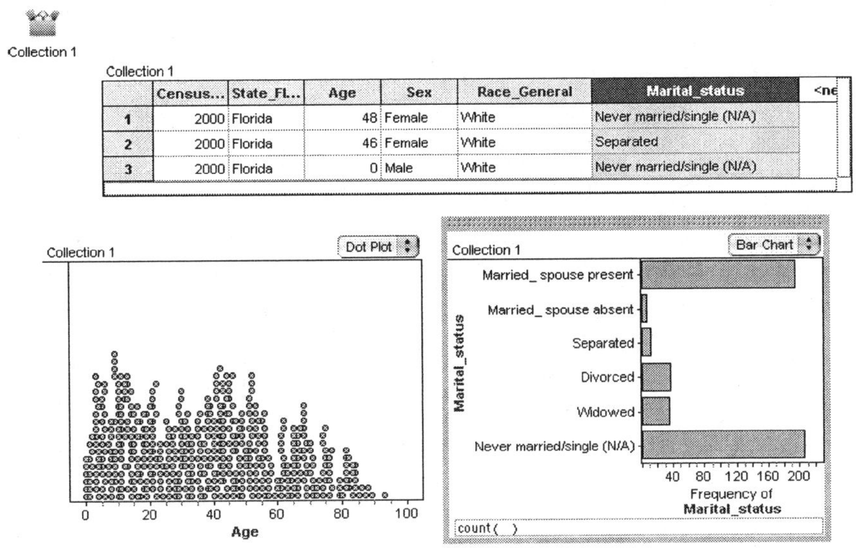

❖ Now for a way-cool thing: with your mouse, click on the bar for all the never-married people. It will turn red (you have selected them all). See what happens in the **Age** graph.

❖ Explore! Make more graphs. Try different things and see what happens. Some ideas:

- Can you find the baby boom?
- Drag **Sex** to the vertical axis of the **Age** graph. What happens?
- Figure out how to rename the collection. (Double-click the name. Call it **People**.)
- See if you can tell whether the majority of widowed people are male or female.
- Figure out how to rescale a graph axis. (Use **Help** if you have trouble.)
- Click on the table to select it, then press DELETE or BACKSPACE. The data don't go away because the *collection* holds the data. Practice making a new table.
- Learn to use **Undo**. It's in the **Edit** menu; its keyboard shortcut is CONTROL-**Z** or **⌘-Z**. **Undo** will be your friend. Fathom supports multiple **Undo**. Likewise, learn about **Redo**.
- What is the sex, race, and marital status of the oldest person? What's the most efficient way to figure that out?
- Talk with other students and share what you have learned.

TWO FATHOM TIPS. First, "maximize" your document. You'll need the screen space. Second, to make a graph, simply drag a new one off the shelf, and label your axes.

Connecting to the textbook

The textbook says, "Any set of data contains information about some group of *individuals*. The information is organized in *variables*."

A computer is not as flexible as a human in figuring out how a data set is organized, so you have to know what the computer needs and get used to the way it does things. For example, in many programs, the individuals appear on the screen as rows in a table, and the variables are in columns. Also, the designers who make computer programs have to decide what to call things—and the names might be different from what you expect.

In Fathom, the standard data table is called a *case table*. Fathom calls the individuals *cases* and it calls the variables *attributes*. This does not affect how things work. The only thing it changes is what words you might look for when you're trying to find a command in a menu.

The textbook also makes a distinction between *categorical* and *quantitative* variables. Fathom calls quantitative variables *numerical*; sometimes you see them called *continuous*. We have made graphs of both types: **Age** is quantitative; **Marital_status** is categorical (each person is in one category with respect to marital status).

Getting Data into Fathom

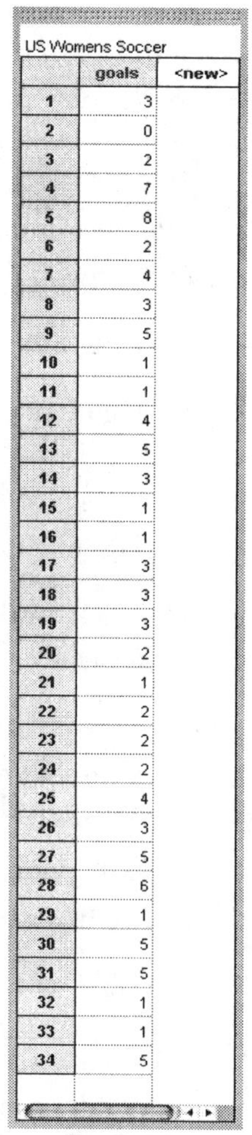
US Womens Soccer

	goals	<new>
1	3	
2	0	
3	2	
4	7	
5	8	
6	2	
7	4	
8	3	
9	5	
10	1	
11	1	
12	4	
13	5	
14	3	
15	1	
16	1	
17	3	
18	3	
19	3	
20	2	
21	1	
22	2	
23	2	
24	2	
25	4	
26	3	
27	5	
28	6	
29	1	
30	5	
31	5	
32	1	
33	1	
34	5	

Let's look at some of the ways you can get data into Fathom.

Example P.7 GOOOOAAAAALLLLLL! (Typing the data in by hand)

As you have seen, you can download Census data from the Internet into Fathom. But you can also just type in your own data.

❖ Open up a new document in Fathom (**File>New**).

In this example in the textbook, we have the number of goals scored by the U.S. Women's Soccer Team in the 34 matches played during the 2004 season. Here are the numbers:

3 0 2 7 8 2 4 3 5 1
1 4 5 3 1 1 3 3 3 2
1 2 2 2 4 3 5 6 1 5
5 1 1 5

The numbers are listed horizontally, but in Fathom, you'll enter them vertically, as in the illustration. Here's how:

❖ Drag a table off the shelf. It will be empty. This is our data table.

❖ We need a column heading for the variable (or as Fathom call it, the *attribute*). Click where it says **<new>**.

❖ Enter **goals** and press ENTER.

❖ Now enter the data. After you type each number, press TAB or ENTER. The cursor will advance to the next row.

❖ Make a graph of **goals**. You can put the attribute name on the vertical or horizontal axis—it doesn't matter. Either way, you get a dot plot. The distribution should resemble the one in your textbook.

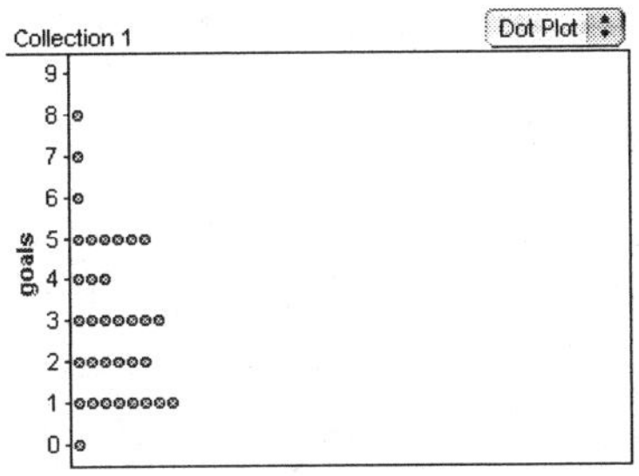

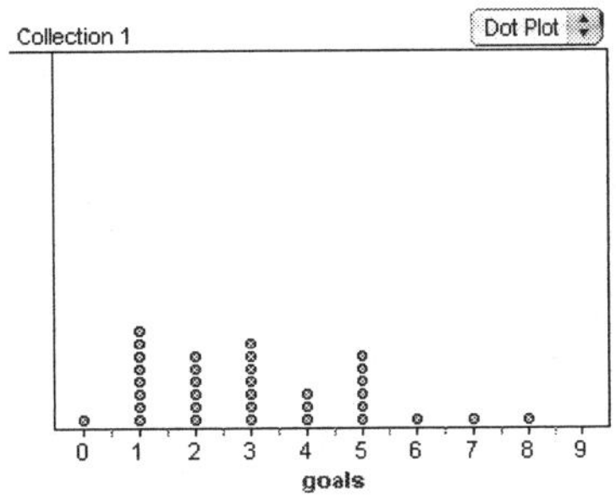

The textbook claims they averaged 3.059 points per game. We can check that in Fathom:

❖ Right-click on the graph (use CONTROL-click if you have a one-button mouse) and choose **Plot Value** from the menu that pops up. The *formula editor* appears.

❖ Enter **mean( )** and click **OK**. The mean appears on the graph—both the number and a line. (We could have entered **mean(goals)**, but **mean( )** with the empty parentheses is sufficient—in fact, even better in this situation. Fathom takes the mean of whatever variable is on the axis.)

CONTEXT MENUS IN FATHOM. Right-click on objects to get *context menus.* We will use them extensively in this guide. If you have a one-button mouse (e.g., on a traditional Macintosh) hold down CONTROL (not COMMAND or ⌘) to get that right-button functionality.

Opening Existing Files

We'll find the next data among Fathom's "sample documents."

The textbook mentions that we "want to know how the percent of students taking the SAT in U.S. states is related to those states' average SAT math scores." This turns out to be a famous relationship, and it's high time you learned what it is.

❖ Choose **File>Open Sample Document....** A dialog box should appear pointing to the Fathom **Sample Documents** folder.

❖ Navigate through the hierarchy to **Social Science>United States>States> States - SAT.ftm**. Open that file. You will see a collection and a table with a lot of columns. The table should look something like the illustration:

States - SAT

	State	V87	M87	V95	M95	V98	M98	V99	M99	V00	M00	V01	M01	Pct...	PctGradsTaking01	<ne
1	Alabama	554	540	565	558	561	555	559	555	559	554	560	559	9	9	
2	Alaska	518	501	521	513	516	514	519	515	514	510	516	519	51	52	
3	Arizona	531	523	525	521	524	525	521	523	523	525	520	523	34	36	
4	Arkansas	554	536	566	550	563	556	563	554	562	550	560	556	6	5	
5	California	500	508	495	511	497	514	497	518	498	517	496	517	51	52	
6	Colorado	537	532	536	538	536	540	534	537	539	542	543	548	31	28	

❖ As in the illustration, drag the edge of the last column heading to the right so you can see the whole variable name: **PctGradsTaking01**. You may need to make the table larger.

❖ That last column is the percentage of high school graduates who took the SAT in each state in 2001. Most of the columns have math or verbal scores for a particular year. For example, **M01** holds the average SAT math score for each state in 2001.

❖ Make a graph with **PctGradsTaking01** on the horizontal axis and **M01** on the vertical. You should see a graph that looks like this:

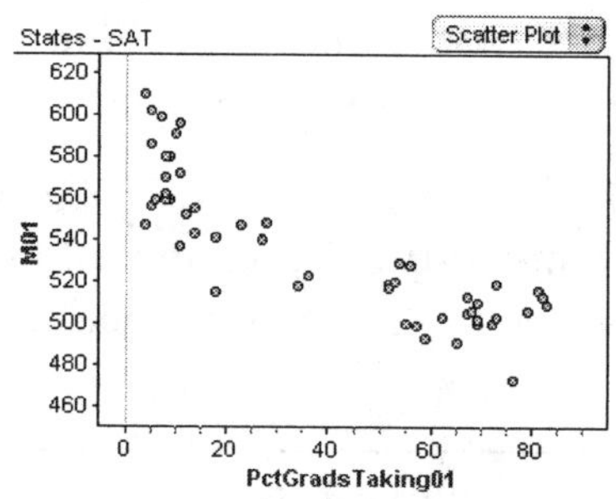

This is a *scatter plot*. It shows the relationship between two quantitative variables. We see that, in general, the higher the percentage of students in a state taking the SAT, the lower that state's average score.

Which state had the highest average score? Point at that data point on the graph and look in the lower-left corner of your window. The name of the state appears in the *status bar*. Useful information appears in the status bar when you point at various objects.

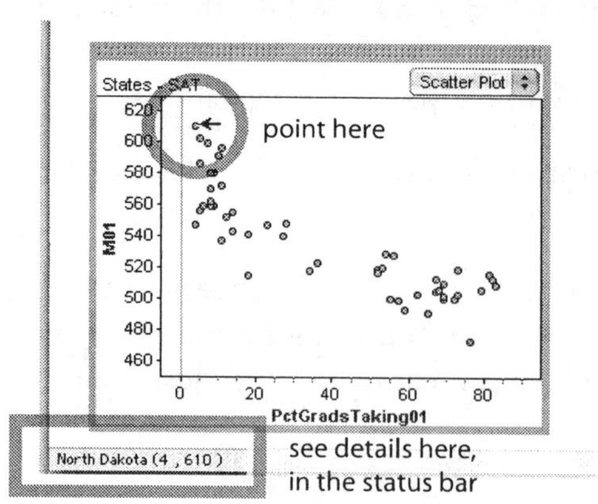

In the illustration, the status bar tells you that North Dakota had the highest math SATs in 2001, an average score of 610. Does that mean that the best math students are in North Dakota? Not necessarily; only 4% of the graduates took the SAT at all.

Dragging data in from Internet sites

On page 30 of the textbook, in the *Web Links* box, you learn that you can find excellent data at **lib.stat.cmu.edu/DASL/**. True. But how do you get it into Fathom?

❖ In a web browser—such as Firefox, Navigator, Safari, or Internet Explorer—surf over to that site, **lib.stat.cmu.edu/DASL/**. This is the Data and Story Library at Carnegie-Mellon University in Pittsburgh, Pennsylvania.

There are many ways to find great data here. Let's do an example:

❖ Click the link to **List all topics**, then **Food**, then **Hot dogs**. There you will see a description of the situation that gives rise to the data.

❖ Click the link that says **Datafile Name: Hot dogs**, which will take you to a page that has the data on it, with a lot of explanatory text.

❖ Now for the magic. Arrange your windows so that the browser is on top, and you can see an empty Fathom window behind it. (You may have to shrink the browser window.)

❖ Grab the little icon (S)that sits at the left of the URL in the browser. Drag it over the blank Fathom document and let go. Fathom will get the data and put it into a collection (which appears). It will look something like this:

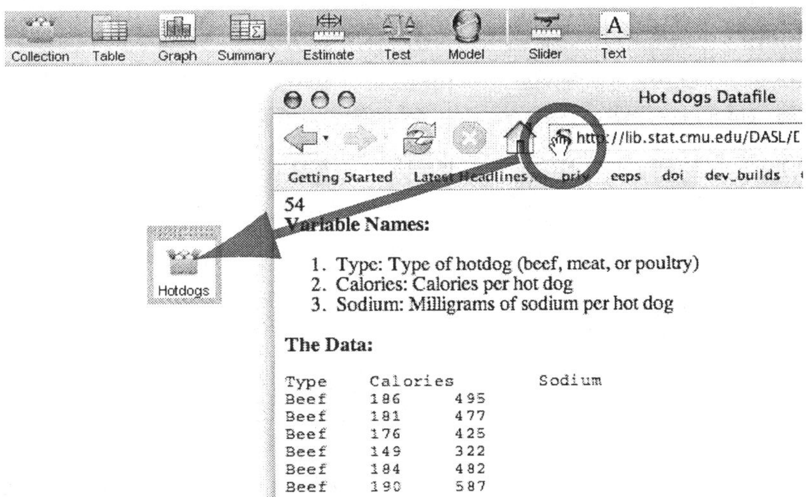

❖ Switch to Fathom, select the collection, and drag a table off the shelf. If all went well, you should see the data.

❖ Just to check, make a graph. Put **Calories** on the horizontal axis, and put **Type** on the vertical. What does that tell you about hot dogs? (And in case it makes you think that poultry dogs are so great, drag **Sodium** to the graph to replace **Calories** and see what you find!)

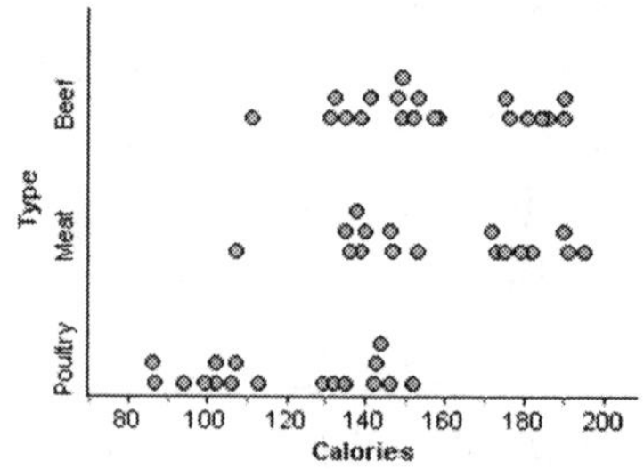

Does this drag-and-drop wizardry work on *all* Internet data? No, but it does work for quite a lot of it, whether the data are in HTML tables, or separated by tabs or commas, or just formatted with spaces. This means that if you can put up pages (in a blog, for example), you can share data this way.

Summary and Conclusion

What should you have learned in Fathom in this chapter?

- ❑ How to start up Fathom and get a fresh document
- ❑ How to get data from a variety of sources: from the Census via the Internet; by typing it in; by opening files from the **Sample Documents** folder on your hard drive or the CD-ROM; or by dragging data in off the Web.
- ❑ How to make data tables and simple graphs
- ❑ How to plot values (such as the mean) on graphs
- ❑ How to get a context menu
- ❑ How to find information on a single point in the status bar

You should also have seen what the main ideas from the preliminary chapter look like when you work with them in Fathom. For example, Fathom handles both quantitative and categorical data, and it has displays that help you look at the *distribution* of your data.

You probably found that, with a little practice, it's really easy to make graphs and explore your data in Fathom.

You probably have questions that begin, "But how do you…?" That's normal, and we will learn a lot more in the coming chapters. We started out easy, doing just a few things with each data set.

Meanwhile, you should explore on your own. You can't break anything! Also, you could skim ahead in this guide or use the online help system (just look in the **Help** menu) to learn more about Fathom.

1 Exploring Data

The previous chapter was an introduction and overview. Now it's time to continue building your statistical skills. From now on, we'll assume you know the Fathom basics: how to make graphs, how to get tables to show your data, and how to get context menus by right-clicking (or CONTROL-clicking) on objects. If you forget, review the preliminary chapter—or simply look in Fathom's **Help** menu.

1.1 Describing Distributions with Graphs

In the previous chapter, you already began to learn about graphs, and how to make them in Fathom. Now you'll learn about more types of graphs and how to make them.

Graphs of categorical variables: Bar charts, ribbon charts, and filters.

We'll begin with the subsection called "Graphs of categorical variables" (page 39). The textbook discusses data about radio station formats. Let's get our own data, using our friend the U.S. Census.[1]

❖ Open a new Fathom document and choose **File>Import>U.S.Census Data...**. A collection appears, and its inspector automatically opens to the **Microdata** panel.

Last time, you downloaded 500 people by pressing the button. This time, before you download, tell Fathom to get each person's educational attainment in addition to his or her sex, age, race, etc. Here's how:

❖ On the left side of that panel, open **Attributes** and then **Person**. Under **Person**, select **Education**. On the right side, check the **Educational Attainment** box. Your screen should look something like the illustration.

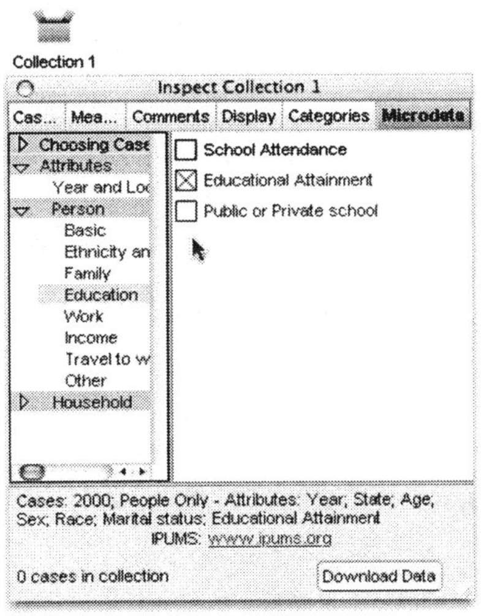

1 If you have no access to the Internet, use the file **01 Educational Attainment.ftm**. It's in the folder called **01 Exploring Data**.

❖ Now click **Download Data**. Fathom will get 500 people, as before, but this time with their educational attainments. Close the inspector.

❖ Display the data in a table. Widen the new "education" column and scroll through the table briefly to get an idea of what's there. Then make the table wide, but short.

MAKING TABLES. If your collection (the box of balls) is selected when you drag a new table off the shelf, the table automatically fills. If not, the table will be empty. In that case, drag the *name* of the collection into the table to fill it with data.

❖ The variable we're interested in is called **Educational_attainment_recode**. That name comes from the Census. While it is good to have variable names that describe the data, this is longer than you need for this purpose. Double-click the name in the table and edit it to be **Education**.

USE DESCRIPTIVE NAMES. Make a habit of creating useful names for variables and collections. In Fathom, double-click names to edit them.

❖ Now make a new graph and put **Education** on the vertical axis. (Horizontal works too; try it and see why we chose the vertical axis.) You get a bar chart.

❖ Useful diversion: Remember how, in the preliminary chapter, we made two graphs, **Age** and **Marital Status**, and then selected different things? Do the same here: Make a second graph, showing **Age**; then select points in the **Age** graph and bars in the **Education** graph. Watch for what lights up.

We are interested in the educational attainment of 25–34-year-olds. Our graph shows everybody. We can get some idea what's going on by selecting the 25–34-year-olds in an **Age** graph, but it's hard to choose exactly the cases we need.

In this case, it works better to use a Fathom *filter*. A filter will restrict the **Education** graph to the ages we want.

❖ Right-click on the graph (CONTROL-click for one-button mice) to get the context menu. Choose **Add Filter**. The formula editor appears.

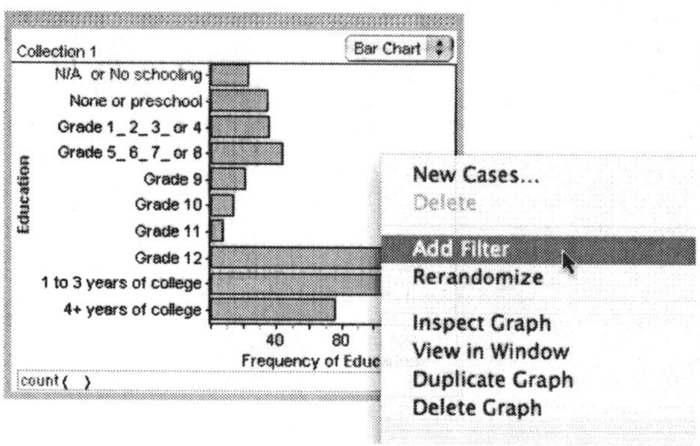

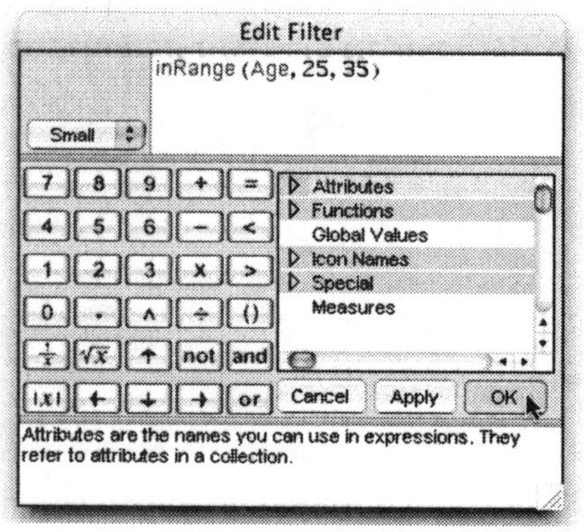

❖ Enter **InRange(Age, 25, 35)**. Click **OK**. The graph will change; now it only shows the **Education**s of the people 25–34. Note: The useful function **InRange** includes the bottom number but excludes the top one—that's why we used 35 in the function and not 34. This is a tradition.

Other expressions work, such as **InRange(Age, 24.5, 34.5)** or **(Age > 24) AND (Age < 35)**. In that case, make **AND** with the keypad on the calculator screen.

Anyway, the textbook states that 29.3% of these young adults have a bachelor's degree. How close is that to the figure for our sample?

❖ Point with the mouse at the bar for 4+ years of college. Look in the lower-left part of the window at the status bar. There you'll see the number and the percentage for that bar. For our sample, it surprisingly close to the national number:

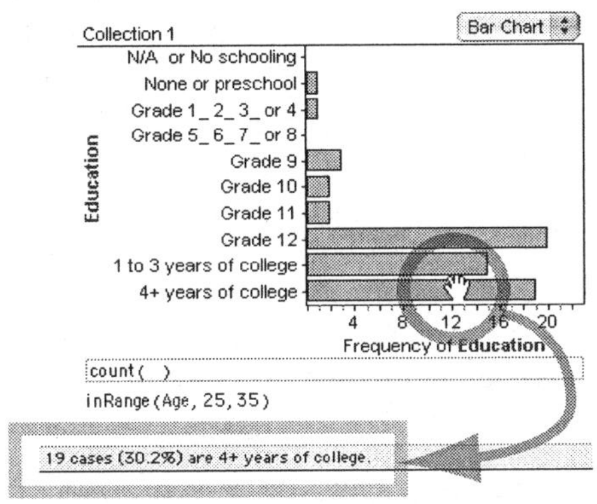

Your graph may not look exactly like this, because you will get a different sample of people.

Use the Status Bar. Point at practically anything and look in the status bar to get useful information. It's easy to forget it's there. Don't.

Next, the textbook mentions *pie charts*. Fathom doesn't make them (as of this edition, anyway). Instead, Fathom makes a *ribbon chart*.

❖ Choose **Ribbon Chart** from the menu in the upper-right corner of the graph. The bar chart changes to a chart where a single rectangle is divided into segments, each representing one category of **Education**. The axis is in percent.

Depending on the size of your graph, the text will overlap for the smaller categories.

❖ You can split a ribbon chart by dragging another categorical variable to the *middle* of the chart. Do that with **Sex**: Drop **Sex** in the middle of the graph—not on an axis—to get a chart like the one below (we show the "drop" in progress):

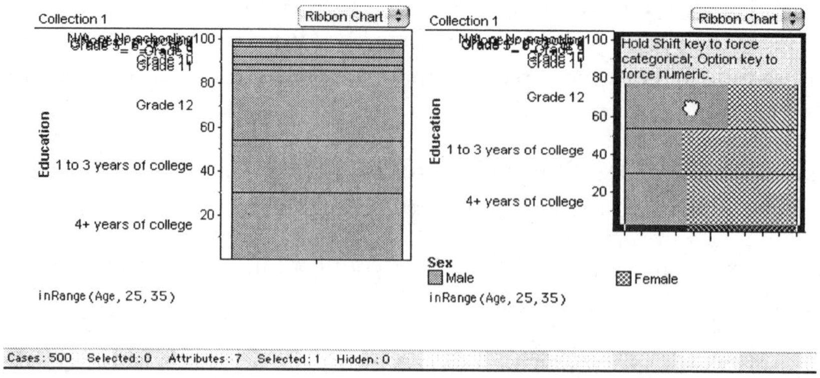

In that graph, you can see that for this sample at least, females make up more than half of the college grads but less than half of those who stopped after high school. Isn't that more interesting than the original graph? Do you think it's true for the whole population or just this sample? Is it true for older age ranges? Those are the kinds of questions you can explore *easily* with technology.

Graphs of quantitative variables: Dot plots and histograms

As we saw in the introductory chapter, when you drop a quantitative variable onto an axis, Fathom makes a *dot plot*. There is one dot for every individual, every case. This is a great way to display data because you can really see where each individual is. If you select an individual in the table, it highlights in the graph, and vice-versa.

In a dot plot, Fathom plots each dot directly over its value on the axis. Sometimes values are so close together that the dots stack in a kind of jagged way. Let's explore that briefly:

❖ Using your Census data, make a dot plot of **Age**. Chances are, the dots will stack in a jumbled way, as in the illustration below left.

❖ Now stretch the graph horizontally (grab the right edge and pull) to give the dots more room. As the graph opens up, the dots will fall down to the axis, as on the right below.

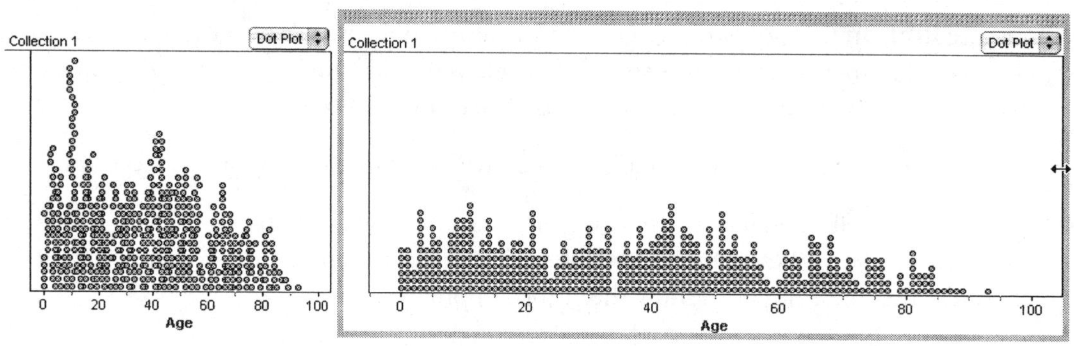

In a dot plot, it's easy, for example, to find the age of the oldest person (93) and to see the overall shape of the distribution. You can also see that the ages of our 500 people seem to be whole numbers; nobody is 17½ years old.

Sometimes there are so many cases, and a dot plot can become so busy, that you want a graph that gives you *less* information so you can see the overall picture. It's time to learn about *histograms.*

❖ Choose **Histogram** from the pop-up menu in the upper-right corner of the graph. The graph changes to a histogram, which resembles a bar chart in that it has bars instead of dots. The height of each bar tells how many people are in that range of ages. The set within that range is called a *class* or a *bin*. The next illustration shows our dot plot beside a histogram of the same data:

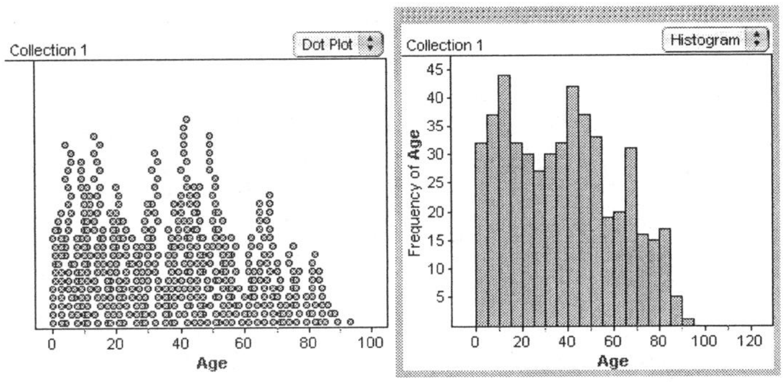

HISTOGRAM TRADITION. A bin or class traditionally includes the bottom value and excludes the top one, so the 15–20 bin includes the 15-year-olds but not the twenties. This is the same convention Fathom uses for the function **InRange( )**.

Grouping the cases together into bins "smooths out" the distribution you see in the dot plot. You no longer have information about individuals (you can't tell the age of the oldest person any more, for example), but you can see the overall shape more easily, and you're unlikely to be distracted by fluctuations. That's because of binning. In this case, Fathom made bins that are five years wide—a convenient number when you're dealing with ages.

Still, why five years? That's a problem with histograms—the apparent shape of the distribution depends on how you set up the bins. In Fathom, you can change bin width dynamically:

❖ Point with the mouse at the boundary between two bins. You'll see a double-arrow cursor. Drag to change the bin width.

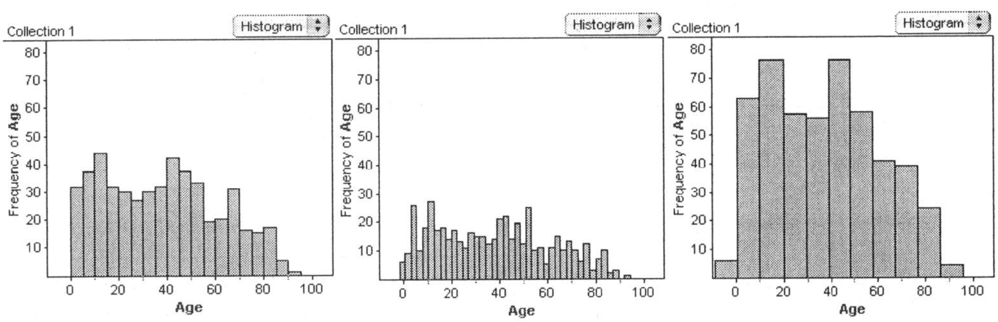

The same data in three different bin widths. Notice how tall the right-hand bars are. If you have fat bins, you will have to adjust the scale to see the tops. (Be sure you understand why fat bins get tall!)

As you play with dragging the bin edges, you might totally mess things up so you can't figure out what's going on. If that happens:

❖ Choose **Histogram** again from the menu in the graph. The histogram's axes rescale so you can see all of the data.

RESETTING GRAPHS. To reset any graph to its "default" scale, choose its type again from the menu in the graph. You can also select the graph and choose **Rescale Graph Axes** from the **Graph** menu or from its context (right-click) menu.

Rescaling will not fix the bin widths if they are in some odd state. Here is how to control the binning precisely:

❖ Double-click the graph to make its inspector appear. Make sure the **Properties** panel (not the **Cases** panel) is showing.

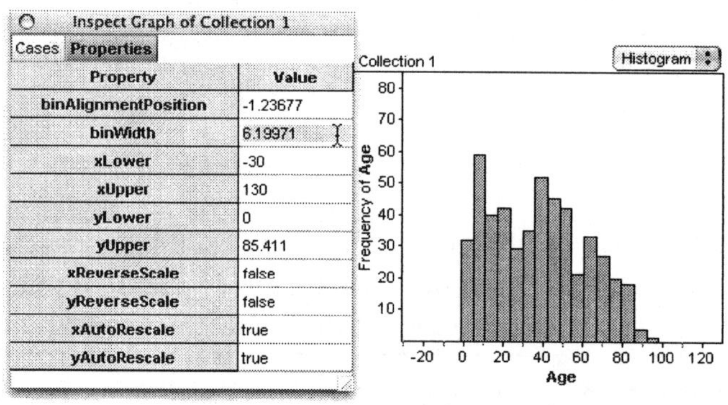

❖ In that panel, edit **binAlignmentPosition** and **binWidth** to change their values, and press ENTER. The graph will update right away. Experiment to figure out how you want your graph to look, and what those two quantities mean.

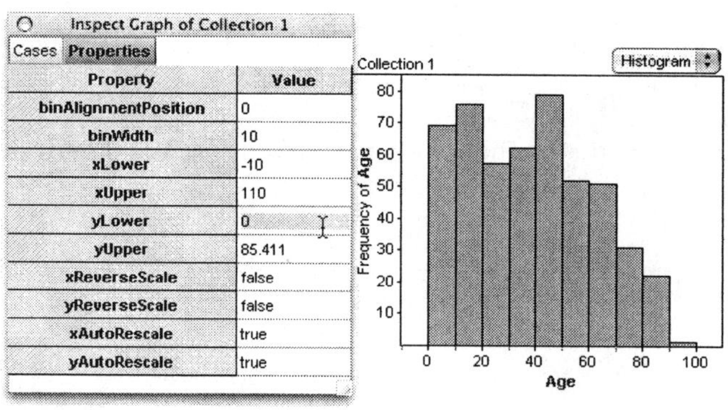

Notice that you can change other things here as well, such as the limits on the graph axes.

THE GRAPH INSPECTOR. Double-click a graph to bring up its inspector. There you can set many properties of a graph. With the inspector, you can tell Fathom *exactly* the scale and bin widths you want. If precision is not so important, you can change axis bounds and bin widths dynamically by dragging parts of the graph.

Try This

In data from the 2000 U.S. Census, you should be able to see the "baby boom"—the result of increased births in the years 1945–60.

- ✎ How old were those people during the Census in 2000?
- ✎ What's the best bin width and alignment to see the boom in your histogram?

The baby boom is not always obvious. If you want a different 500 people, open the collection's inspector to the **Microdata** panel and press **Download Data** again. Fathom will replace the data with all new people. If you want *more* people, open **Choosing Cases** in that inspector panel and select **Sample Size** to specify how many people you want.

Stem Plots

Fathom does not make stem plots. Everything you may want to do with a stem plot you can do with a Fathom dot plot. And the dot plot does not round off, nor does it impose any particular binning, as stem plots do.

Why study stem plots at all, then? Because stem plots are one of the best ways to get an idea of the shape of a distribution *by hand*. If you're analyzing your data with a decent data-analysis program, you don't need a stem plot.

Ogive Plots

On pages 60–62, the textbook describes *ogive plots*. In Fathom, they're called *percentile plots*.

- ❖ Open **EX01_11 Presidential Ages.ftm**. Make a graph with **age** on the horizontal axis. You will see a dot plot showing the distribution.
- ❖ Choose **Percentile Plot** from the pop-up menu in the corner of the graph. A new plot appears (it should look like the illustration) that corresponds to Figure 1.12 on page 61 of the textbook.

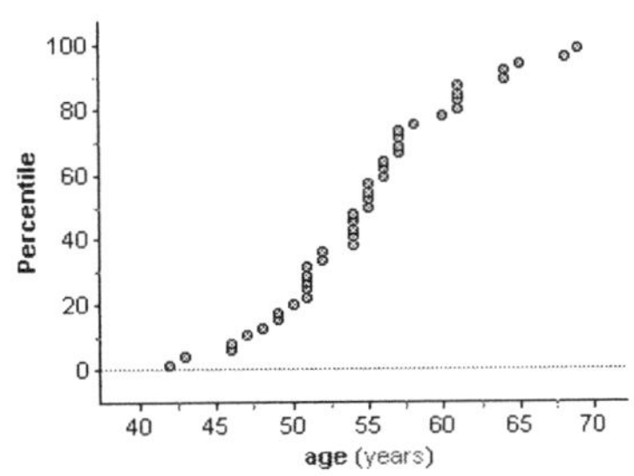

Time Plots

We'll talk about these in more detail in Chapter 3 (see the box on page 49 in this guide), when we study data about manatees. In the meantime, try this:

- ❖ Open **01 Nebraska Gas.ftm** and make a new, empty graph.
- ❖ Put **Price** on the *vertical* axis. You get a dot plot.

- ❖ Use the menu in the corner of the graph to change this to a **Line Plot**. You'll see the pattern in the prices not unlike Figure 1.14 in the text. Note that the dots are connected.

- ❖ By default, the horizontal axis is just the case number, which is not too informative. If you're using Fathom 2.1 or later, drag **Date** to the horizontal axis (otherwise drag **DecimalYear**). You should see a graph like the one in the illustration.

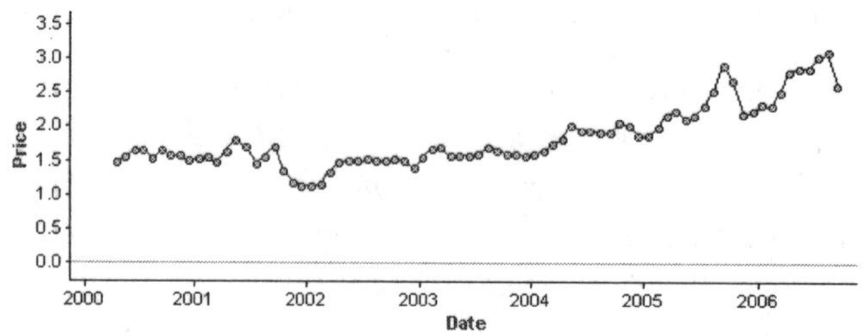

1.2 Describing Distributions with Numbers

You can tell a lot about a data set from a graph or a table, but sometimes you need to summarize the data with numbers. For example, if you're making cheesecake in a factory, you might have the weights of the last 137 cheesecakes that came off your assembly line. You don't need all of the weights for your report to the Big Cheese, but you might need the *average* weight. And you might want to know how much the cheesecakes *varied.*

A number that tells you about a "typical" value is a *measure of center.* A number that tells you about variation is a *measure of spread.*

There are several different ways to get numerical summaries in Fathom.

Summary Tables

We took another sample of 500 people from the 2000 U.S. Census. Among those 500 people, 43 were high-school graduates who were currently attending school. You can see the data in **01 HSG attending School.ftm**.

- ❖ Open the file. It is in the folder for Chapter 1, which is called **01 Exploring Data**.

- ❖ Make a graph of **Age**. It will look like the dot plot below, though your plot may be scaled differently:

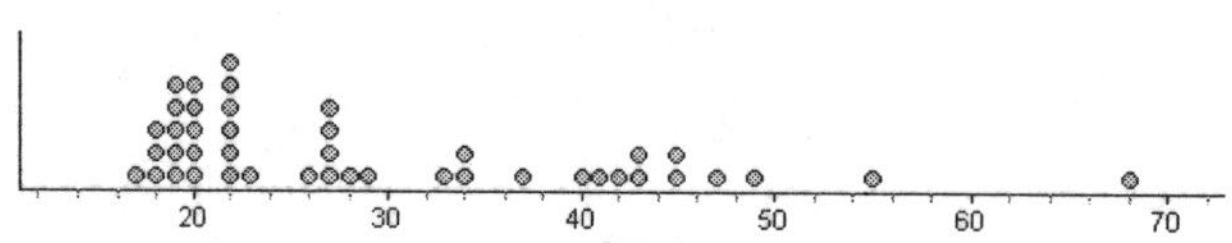

What is their average age? Use a Fathom *summary table* to find out:

- ❖ Drag a new summary table off the shelf. 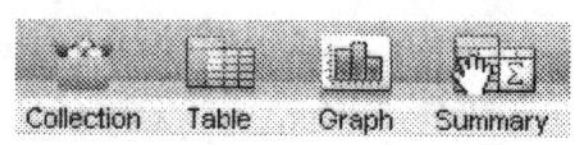

- ❖ Drag the name **Age** from the case table to a column or row heading in the summary table. A box will highlight when you're over a good spot.

The summary table shows the mean by default:

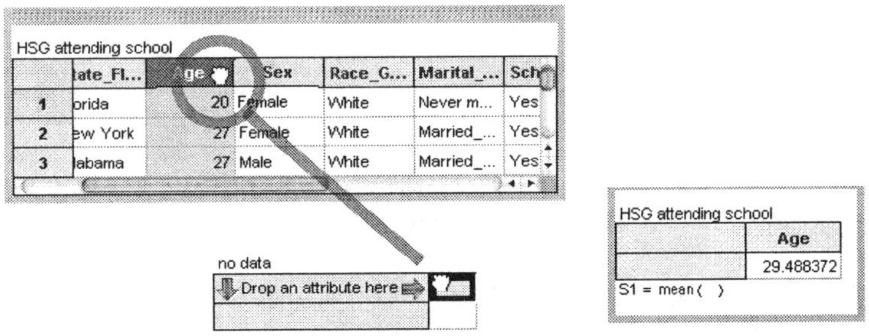

You can get more than just the mean, of course. Right-click on the summary table (use CONTROL-click on the Mac) to get the table's context menu. Choose **Add Five-Number Summary** from that menu and see what happens:

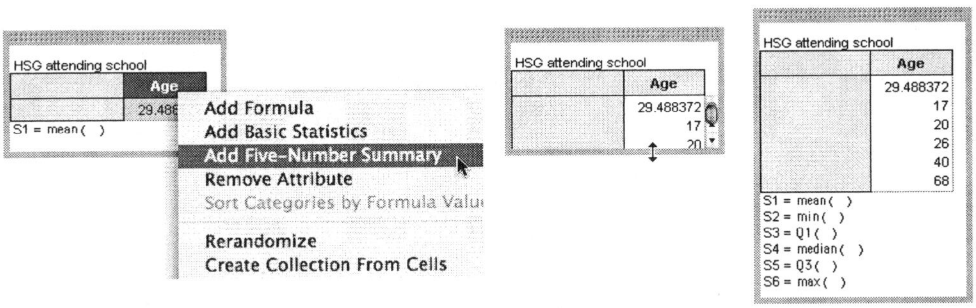

You will need to stretch the summary table vertically in order to see all of your results. Now you have a number of other measures besides the mean: the minimum, the first and third quartiles (labeled **Q1** and **Q3**), the median, and the maximum.

The numbers in the box are in the same order as the labels down below. If you get confused, just click on one of the labels, and the corresponding number will highlight. Try that: Click on the label that says **S4 = median( )** and see that the median is 26.

❖ To add even more numbers to our statistical summary, right-click again on the table and choose **Add Basic Statistics.**

Now we have added the count, the standard deviation (**stdDev**), the standard error (**stdError**), and the number of missing data points.

Those labels at the bottom of the summary table are important. They're not just labels—they're actually formulas. You can put *anything* in one of those formulas, not just what Fathom gives you by default. For example:

❖ Right-click the summary table and choose **Add Formula** from the context menu. Fathom's formula editor appears.

❖ In the editor, enter **mean( ) + 1**. Click the **OK** button to close the editor.

You should see the new line, in which Fathom has calculated the mean plus one. To edit an existing formula, double-click it.

How do you get rid of a line in the summary table? Right-click on the formula for the line (the label at the bottom) and choose **Clear Formula** from the context menu.

Plotted Values

You can also use *plotted values* to calculate summary statistics using a Fathom graph. We'll keep using our age data in **HSG attending School.ftm**:

❖ If you haven't made a graph of **Age** yet, do so now. You should get a dot plot. Stretch it out so you can see the individual stacks of points (like in the illustration).

❖ Right-click on the graph (CONTROL-click for Mac users) and choose **Plot Value** from the context menu.

❖ In the formula editor, enter **mean( )**. Just leave the parentheses empty. Close the editor with **OK**.

Now you see the value for the mean plotted on the graph, and the numerical value below.

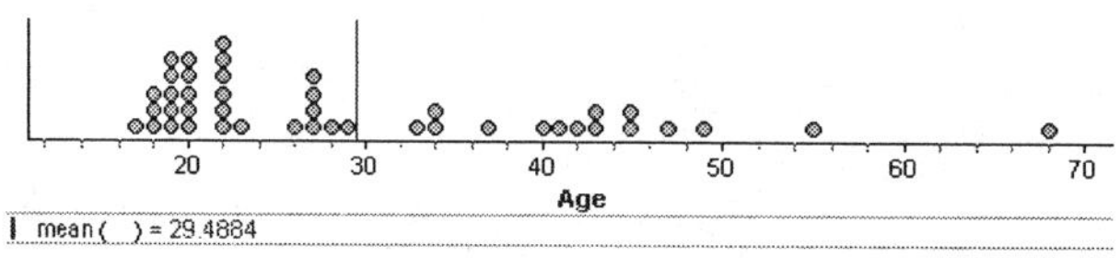

Let's plot the median as well:

❖ Again right-click on the graph and choose **Plot Value** to bring up the formula editor.

❖ This time, enter **median( )** and close the editor with **OK**.

Both lines appear. You can tell which is which by the color coding. Just as in the summary table, double-click the formula to edit it. Also, **Clear** the formula to eliminate the line altogether.

Box Plots

Another way to get summary information on a graph is to use a special-purpose plot. One of these is the *box plot*. To make a box plot,

❖ Choose **Box Plot** from the pop-up menu in the corner of the graph. The illustration shows a box plot and a dot plot side by side:

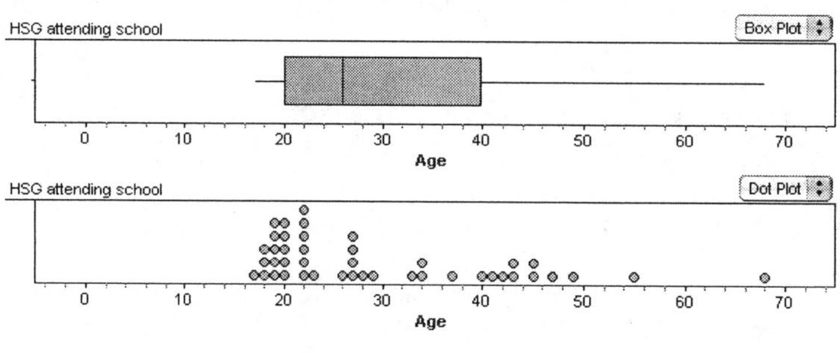

(If you had the mean plotted on your dot plot, it will still be on the box plot.)

The box plot gives you a terrific visual impression of a distribution. In Fathom, you can get numerical results as well. Point with the mouse at one of the box plot's boxes and look at the lower-left corner of Fathom's window. There, in the status bar, you'll see the actual values of the box's dimensions. The same goes for the whiskers.

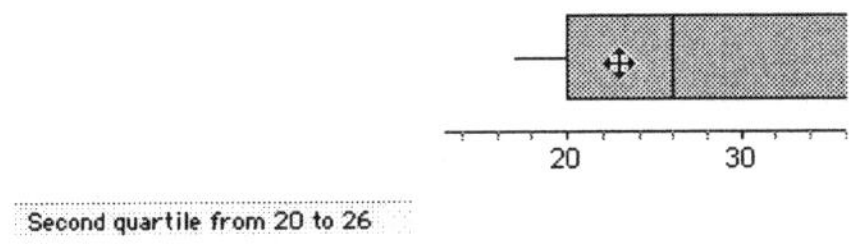

USE THE STATUS BAR FOR DETAILED INFORMATION FROM PLOTS. Point at any part of a plot—a bar, a box, a whisker, or even a single point—to get more information from the status bar.

Side-by-Side Box Plots

Box plots are really good for helping you compare two groups. Still using our data about the ages of high-school graduates (**HSG Attending School.ftm**), we might wonder whether there is any difference between women and men.

To see this in a box plot, drag the relevant attribute to the *other* axis. In this case, the relevant attribute is **Sex**.

❖ Make a box plot of age by dragging **Age** to the horizontal axis and choosing **Box Plot** from the popup menu in the corner of the graph.

❖ Drag the name **Sex** from the table to the vertical axis of your box plot.

The plot splits, showing separate box plots for the men and women.

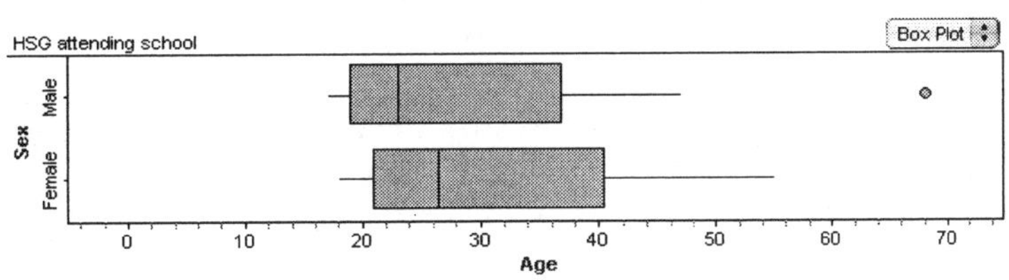

You can see that (except for one older man) the women seem to be a little older than the men in this sample.

FATHOM MAKES MODIFIED BOX PLOTS. As you can see in the illustration, Fathom's box plots show outliers. Fathom uses the 1.5 IQR rule as described in the textbook.

Splitting a Summary Table

You can also compare groups quantitatively in a summary table:

❖ Make a summary table showing the mean of **Age** in **01 HSG Attending School.ftm** by dragging **Age** to the left-hand box in the table (i.e., make **Age** a row heading).

❖ Drag **Sex** onto the table, making it a column heading.

❖ Expand the table as necessary to see all of your results.

You will see the means for each sex and the mean of the whole sample. The next row holds the counts, showing how many cases are in each table cell.

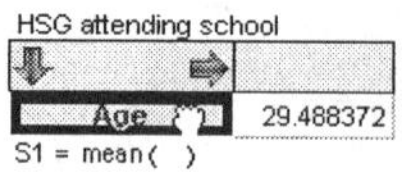

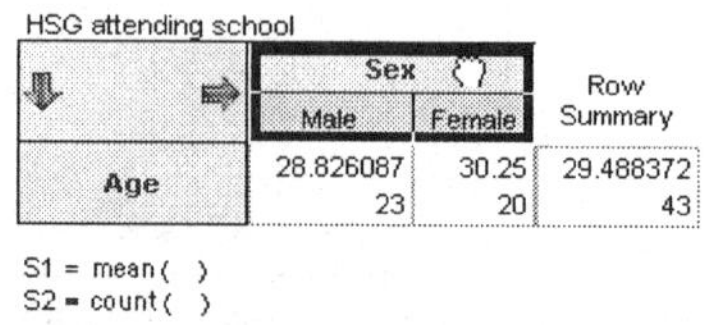

You can add additional statistics to this table as well, using the same techniques we discussed earlier under *Summary Tables.*

Exercises

1.33[2] SSHA Scores

❖ Open the file **EX01_33 SSHA Scores.ftm**. The file will be in the folder for the chapter, in this case, **01 Exploring Data**.

(a) To make a side-by-side box plot when the data are organized this way (one column for men, one for women), put both attributes on the same axis. To do that, when you drag the second variable in, drop it onto the plus sign that appears.

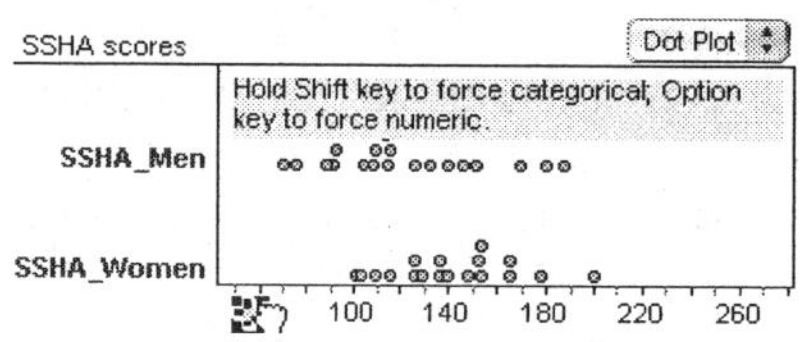

(b) Use **Plot Value** to plot the **mean( )** on the graph. You can see both means.

Make a summary table to compute the two means. You will need to drop the second attribute on the arrow that points to the first.

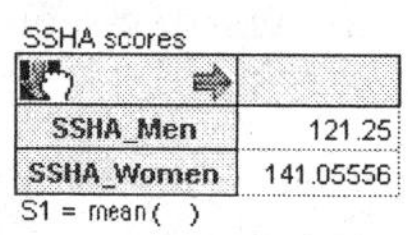

Add other statistics (such as the standard deviation) to the summary table in order to get the numerical summaries.

(c) Write a paragraph comparing the SSHA scores for men and women.

2 This corresponds to Exercise 1.33 in the textbook. The "lettered" parts refer to the corresponding parts of the exercise.

Stacking Attributes

Fathom will let you organize the data in another way. Let's continue with the SSHA data from Exercise 1.33:

❖ Select the collection (the box of gold balls) and choose **Stack Attributes** from the **Collection** menu. A new collection appears.

❖ Make a case table for the new collection by dragging one off the shelf.

❖ Compare the two case tables. How are they the same? How are they different?

❖ Make side-by-side boxplots from the new collection by dragging one attribute to one axis and the other to the other axis.

Stacked SSHA scores

	Group		Row Summary
	SSHA_Men	SSHA_Women	
Value	121.25 20	141.05556 18	130.63158 38

S1 = mean( )
S2 = count( )

❖ Similarly, plot the **mean( )** and make a summary table comparing men's and women's SSHA scores.

❖ Compare the summary graphs and numbers.

✎ Which way of organizing data do you prefer? Why?

WHICH WAY IS BEST? Although many prefer the first way (the tables are smaller, for example) the second way is ultimately easier to use for statistics. Why? Because each row in the table represents a single observation. In the original data, each row contains a man's and a woman's score—yet those scores are not really connected with each other. If we were investigating the study habits of siblings, that would be a different story.

More exercises

1.35: INTERQUARTILE RANGE.

❖ Open **01 HSG Attending School.ftm** again.

❖ Make a box plot and a dot plot of **Age**. Remember that the interquartile range is the total width of the box in the box plot.

❖ Use **Plot Value** to plot the **mean( )** on the box plot.

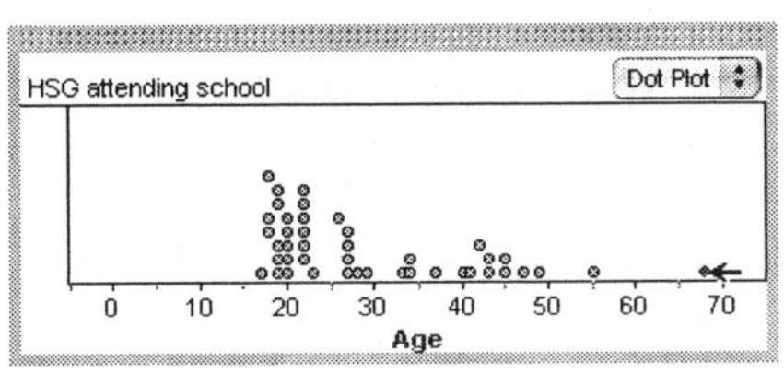

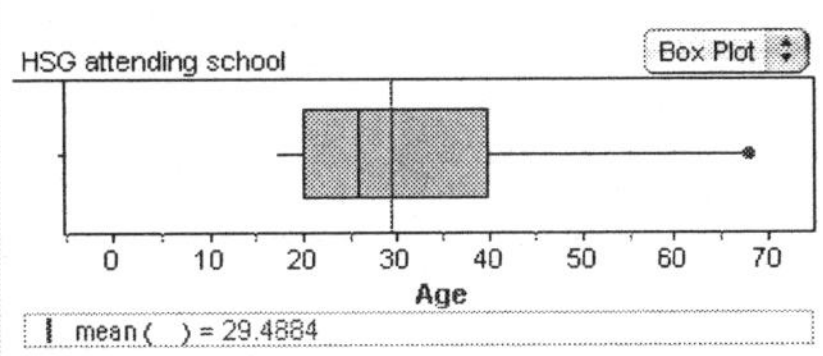

❖ Now use your mouse to grab the oldest person on the dot plot. Drag it left, making that old guy younger. Watch the mean, median, and interquartile range.

❖ How do they change as you drag the old guy? Which changes more? Why?

❖ Extension: Figure out a similar way to compare IQR and standard deviation in terms of how much they change when you drag points around.

❖ Make sure you know how to tell that it's an old *guy*.

1.39 PHOSPHATE LEVELS.

❖ Enter these data:

5.6 5.2 4.6 4.9 5.7 6.4

into a new Fathom table, as values for the variable **Phosphate**. We're going to compute the mean and standard deviation in a new way.

❖ Make a dot plot of the data.

❖ Choose **Add Movable Line** from the graph's context menu. Drag the line to see how it works.

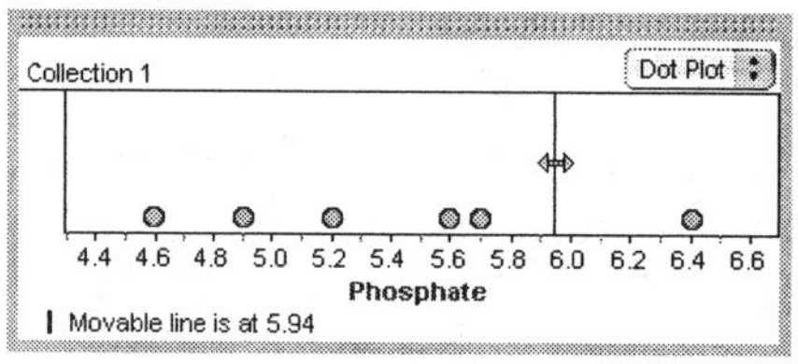

❖ Choose **Show Squares** from the same menu. Six squares appear, and the sum of their areas appears at the bottom of the graph. The squares are built so that the side of each square is the distance from a point to the movable line.

(a) Drag the movable line to see what happens. If the line is far to one extreme, the squares are quite large. Find the spot where the sum of the squares is a *minimum*. Where is the movable line? What is the minimum sum of squares?

(b) The sum of squares is the sum of the squares of the deviations. Divide it by five and take the square root. What number do you get?

❖ Make a summary table of **Phosphate**, and find all its basic statistics.

✐ What is the significance of the value of the movable line you got in (a)?

✐ What is the significance of the number you got in (b)?

✐ We're using the movable line to test out possible "typical" values. Why would we want a typical value to have the minimum sum of squares?

✐ Explain why the sum of squares in (a) is the sum of the squares of the deviations.

1.45 Raising Teachers' Pay I. After you answer the questions in the textbook, test out your ideas by inventing some teacher salaries between $30,000 and $60,000. Enter five or more of them into a new Fathom case table as **salary1**.

❖ Compute the mean, the median, the standard deviation, and the interquartile range. Use a summary table; one formula for interquartile range is **IQR( )**. (See the figure.)

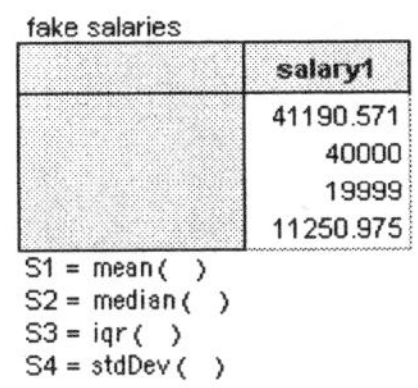

❖ Create a new column (a new attribute) named **salary2**.

❖ Right-click on the *name* **salary2**, and choose **Edit Formula** from the context menu. The formula editor opens, saying "**salary2 =** ."

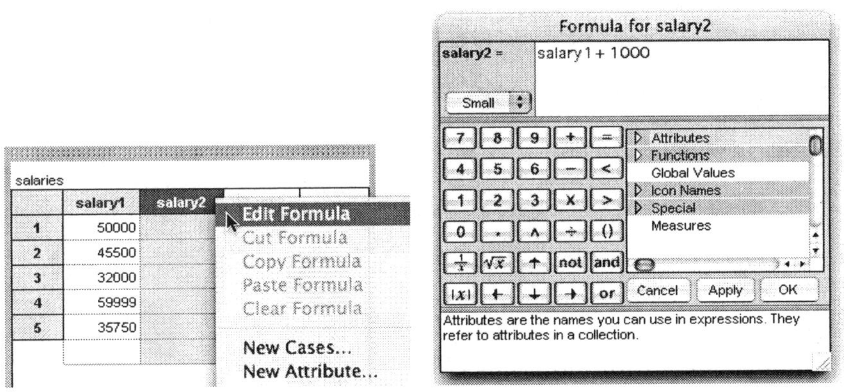

❖ Enter **salary1 + 1000**. Click **OK**. You just gave each teacher a $1000 raise.

❖ Put **salary2** on the same summary table as **salary1** (dropping it on the arrow that points to **salary1**, otherwise it will replace that attribute).

✐ Explain the relationships between the statistics for **salary1** and those for **salary2**.

✐ How will those relationships change if you change some of the salary numbers you invented? Try it and verify your prediction.

Computed Columns. You just learned the basics of how to make a computed column in Fathom: Right-click the attribute name, choose **Edit Formula**, and enter the formula into the formula editor.

1.46 RAISING TEACHERS' PAY II. After you answer the questions in the textbook, test out your ideas in the same way as in Exercise 1.45, above.

- ❖ Make a new attribute, **salary3**, with a formula that gives each teacher a 5% raise over **salary1**.

  - ✎ Explain the relationships between the statistics for **salary1** and those for **salary3**.

  - ✎ Explain why the relationships are different from those in Exercise 1.45.

1.48 GET YOUR HOT DOGS HERE! Get the data off the Internet from the DASL site. See pages 11–12 in this guide to learn how.

You can do many of the remaining exercises using Fathom. If the exercise needs data that's in the textbook, chances are excellent that there is a Fathom file with the data already in it.

Sometimes you will have to approach things a little differently if the textbook supplies summary statistics instead of individual data. For example,

1.49 WHO MAKES MORE? Use the file **EX01_49 Who makes more.ftm**. We do not have the salaries of the individuals, only the number that are in each salary range.

- ❖ Make a graph and put **salary_K** on one axis and **pWomen** on the other. You get a plot that shows the proportion of women in each salary category. It's not a histogram, but it serves.

- ❖ Select the graph and **Duplicate** it (CONTROL-D or ⌘-D).

- ❖ Drag **pMen** to the same axis as **pWomen** in the new graph, replacing **pWomen**.

Now you can compare the graphs and answer the questions in the textbook.

2 Describing Location in a Distribution

In Chapter 1, you learned how to use Fathom to help you explore data, especially how to look at *distributions* of data. In this chapter, we look more deeply. The overarching idea is to develop tools that will help you compare a data point to the distribution it's in. To that end, you will learn about standard scores, also known as *z-scores*, and about percentiles. Then we'll learn about the Normal distribution.

2.1 Measures of Relative Standing and Density Curves

In the textbook, at the very beginning of this section, you see 25 scores on a test from Mr. Pryor's class:

79	81	80	77	73	83	74	93	78	80
75	67	73	77	83	86	90	79	85	83
89	84	82	77	72					

We'll use these data to show you how to calculate *z*-scores and percentiles in Fathom. We'll also use this as an opportunity to review some Fathom techniques.

❖ The data are in the file **02 JennyScore.ftm** which is located in the folder named **02 Describing Location**.

❖ Make a dot plot of the data. See how it resembles the stem plot in the textbook, and how it differs.

❖ Make a summary table and use it to get **Basic Statistics** and the **Five Number Summary**. Your screen should have objects like these:

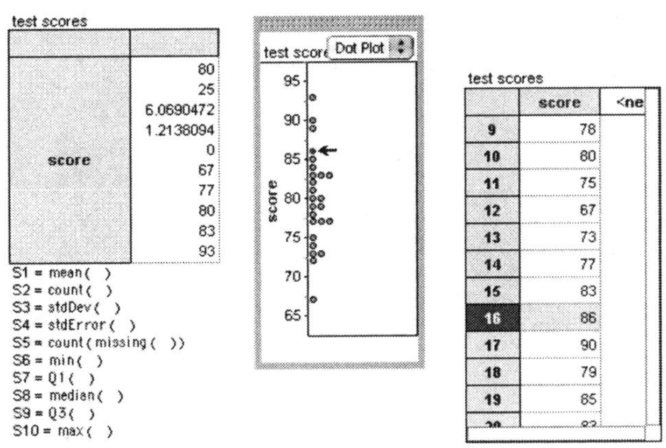

Notice how the values for the statistics are mostly the same as in the textbook. We have selected Jenny's score (86) in the table; it highlights in the dot plot.

What's not the same as in the textbook, and why

The textbook shows some Minitab output. The first and third quartiles are different in Fathom. How is that possible? Statisticians have different conventions about how to calculate these things (and percentiles too, which we will get to shortly). Here are just two of the conventions for calculating the first quartile value, **Q1**, using data from this data set:

1. Find the median of the values that are below the median of the whole data set (80), *including the median*. That makes 13 values; the median of these is 77.
2. Find the median of the values below the median (80), *not including the median*. That makes 12 values, and their median is 76.

Which is right? It doesn't matter. The important thing is the overall shape of the distribution. Whether the first quartile is 76 or 77 is not going to change your mind—the distribution is centered around 80 and roughly symmetrical. If you're comparing two data sets, use the same convention on each. And if you're comparing your analysis with somebody else's, just remember that if there's a discrepancy, it may be because your software uses a different convention from theirs.

Onward!

Now we want to get Jenny's standard score, her *z*-score. We'll calculate everybody's.

❖ Make a new variable, **z**, in the data table. Right-click on the column heading and choose **Edit Formula** from the context menu. Give it a formula that computes the standard score, such as:

z = (score – mean(score)) / stdDev(score)

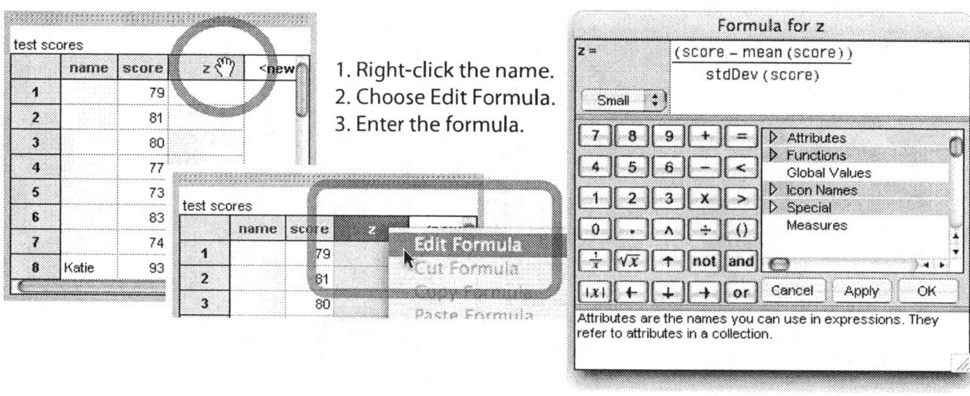

(Don't type "**z =**"; that's in the editor already.) The function **stdDev( )** computes standard deviation. Capitalization does not matter.

Reminder: Making Computed Columns. To set values by formula, you need the formula editor. Right-click on the name of the variable (**z**) and choose **Edit Formula** from the context menu. You will use this often, so it's worth remembering.

The textbook explains how we know that her score is about one standard deviation above the mean; your table should show Jenny's *z*-score as 0.9886. That's about one. Good.

✐ We know the mean and standard deviation. So we could have just written

z = (score – 80) / 6.07

It's shorter, and easier to type, but the longer formula is better. Why? For example, if you add another **score** to the data set, this short formula will no longer be correct, but the longer one is more general.

Example 2.3: Percentiles

The textbook explains that Norman scored at the 8th percentile on the test because 8% of the scores were at or below his score of 72. Let's use Fathom to compute these percentiles:

❖ Make a new variable, **percentile**, and give it the formula

percentile = 100 * rank(score) / count( )

The function **rank( )** gives the "rank" of that case counting up from the bottom. Norman is second from the bottom, so his rank is 2. The function **count( )** returns the number of cases, in this case, 25. The number **100** converts the decimal into a percentage: $100 * 2 \div 25 = 8$.

So Norman is in the 8th percentile, just like in the textbook. Now notice that the two students with scores of 80 are in the 54th percentile. Refer to the textbook (pp. 119–120) and see that according to its method, they should be in the 56th. What gives?

Again, it's a difference of convention. It doesn't really matter as long as you're consistent. As the textbook says, "Calculating percentiles is not an exact science!"

✐ Figure out why there is a discrepancy between the textbook's percentiles and Fathom's, and why they are sometimes the same and sometimes different. Hint: Investigate Fathom's **rank( )** function.

Exploring *z*-scores

The steps below will help you understand *z*-scores better, and teach you a great shortcut!

❖ In a new Fathom document, make a new table with one attribute (variable), named **x**. Give **x** five values, 1, 2, 3, 4, and 5.

❖ Make another variable, **z**, and make it the standard score of **x**. Instead of using the formula we used above, with **mean** and **stdDev**, we'll use a shortcut:

z = zscore(x)

- Make a dot plot of **x**. Make a separate dot plot of **z**. Your graphs should look something like this:

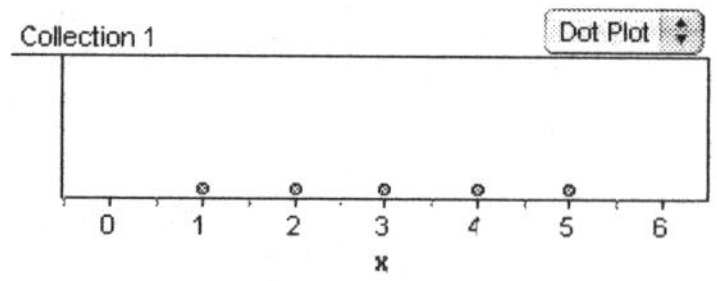

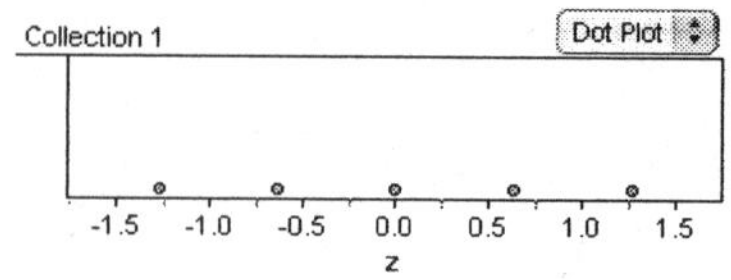

- Make sure you understand what you see so far, for example, why some of the **z**'s are negative but the **x**'s are not.

- Now for the cool part. In the **x** graph, drag the points, one at a time, towards zero, starting with the point at 1. Watch what happens in the **z** graph. See if you can explain why the "other" points get farther away.

- By the time you're ready to drag 5, it will be off the **z** graph. As you drag 5 in, watch what happens to the **z** points.

This is a very strange thing to see the first time. Stick with it, and see if you can understand how the parts of the formula for the *z*-score interact to produce the behavior you see as you drag the points. When you can explain it clearly to your instructor, insist on extra credit.

Chebyshev's Inequality

We're now going to use some of the same machinery we just used—particularly the idea of dragging points around—to demonstrate how Chebyshev's Inequality works.

The idea is that for any distribution, no more than $1/n^2$ of the data points will be more than n standard deviations away. What does that actually mean? Let's explore it in Fathom.

- Open the file **02 Chebyshev.ftm**. You'll see a pretty busy document that looks like the illustration on the next page.

The collection called **Chebyshev** contains 40 points, each with a value of **x** and a corresponding **z**. These are plotted, each variable in its own dot plot, below the table.

At the top, in the middle, is a *slider* called **n**. We use it to specify how many standard deviations we're studying. The vertical lines in the **x** graph show the mean plus or minus **n** standard deviations. The question is, for a given value of **n**, how many data points can we drag outside the lines? Let's start with **n = 2**.

- Set **n** to **2** in the slider (you can drag it there, but typing in the value is more precise). No points are outside the lines right now. The "computed" slider **nout** shows how many are outside.

- Now drag one point in the **x** graph to the right until it is outside the lines. Notice how **nout** increases to 1.

- Keep dragging points, trying to get as many outside the lines as you can.

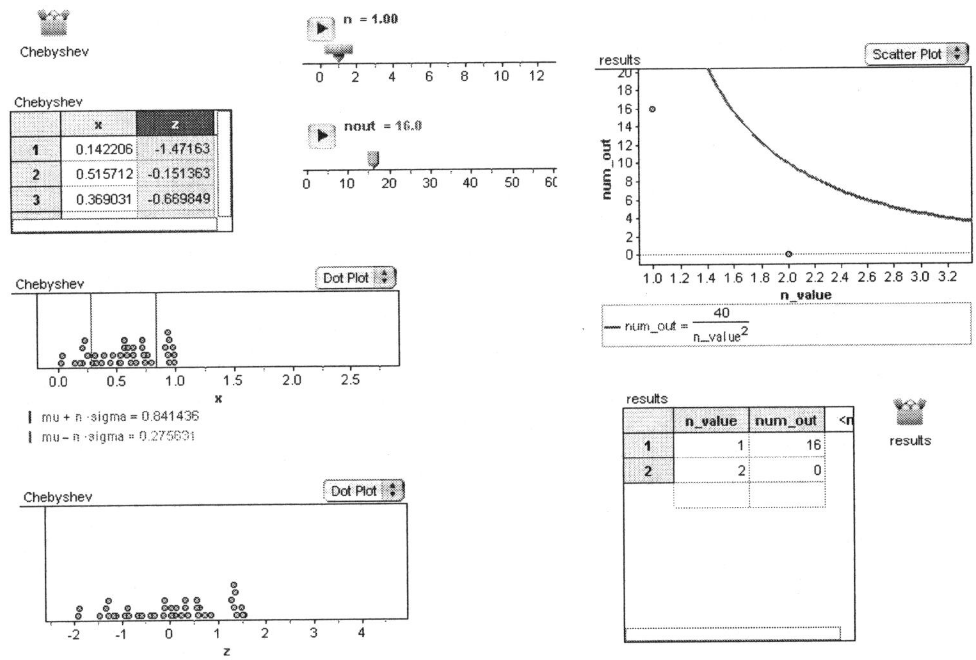

- ❖ When you get as many points out as you can, record that number of points (**nout**) in the **results** table on the right. Type in **2** for **n_value** and the number you got outside the lines under **num_out** (we have already recorded two points; you can do better).

Chebyshev's inequality states that you cannot get more than ¼ of the points outside 2 standard deviations. That would be 10 points. If you got 8 out, you're doing really well. The graph has a curve on it, **num_out** = $40/n^2$, so you can see how close you're getting to the limit. (The 40 is the number of points, so $40/n^2$ is the limiting *number* corresponding to a *proportion* of $1/n^2$.)

- ❖ When you've done what you can with 2 standard deviations, change the **n** slider and repeat the process. See how many points you can get outside the lines for different values of **n**, and record those results in the **results** table.

You will discover that Chebyshev—smart guy that he was—was right. (If you find a way around him, you will become a famous mathematician almost instantly.)

What's the point of doing this? Really, there are two points: first, to give you a chance to see how it's possible that a strange assertion like Chebyshev's could be true for any distribution at all. Second—and more important—doing this should give you a better feel for how standard deviations work.

Here is an insight: To get more points outside 2 standard deviations, that is, make more points with a *z*-score greater than 2, we can use two strategies. The obvious one is to move a few points farther out (increasing the *z*-score of those points). Or we can be subtle and move many points *in*. This *decreases* the standard deviation of the whole data set, thereby *increasing* the *z*-scores of the ones that aren't moving.

As you have probably seen, the first strategy works at first. When you move the first point out, it works fine. It's easy to make one exceptional point. But when you keep doing that, pretty soon the "extreme" points have too much company—they aren't exceptional any more.

You can make them more exceptional relative to the rest by making the rest of the points as identical as you can (for example, stop dragging and just type in a bunch of zeros). The more the masses conform, the less you have to do to be outrageous. One final hint: To maximize the number of outliers, it helps to have outrageous cases on both sides of the mediocre middle.

Exercises

2.5 UNEMPLOYMENT IN THE STATES. Open the file **EX02_05 Unemployment.ftm**. Instead of May 2005, this file has data from January 2006. Use Fathom to answer the same questions as are in the textbook. Some hints and extensions:

- Compute percentiles and *z*-scores for all the states, as we did for the test scores earlier.
- In Fathom, you can sort a table. Right-click on an attribute label (a column heading) and look at the context menu. Choose **Sort Ascending** or **Sort Descending**.
- To answer part (e), don't do fancy calculations. Just sort by *z*-score and count the states in the list that have, for example, $z > 2$ and $z < -2$. Since there are 50 states, it's easy to compute proportions.
- Make a box plot as well. Identify the highest-unemployment state. That state also had the greatest *increase* in unemployment in the U.S. in the preceding year (i.e., during 2005). Make a conjecture about why that state was having employment problems.

Example 2.4 From histogram to density curve (extension)

Let's use Fathom to explore this idea of density a little more.

- Open the file **EG02_04b heights.ftm**. It has the heights (and ages, sexes, and foot sizes) of 100 people.
- Make a histogram of **height**.
- With the graph selected, choose **Object>Duplicate Graph** or use the keyboard shortcut CONTROL-**D** (⌘-**D** on the Mac). Drag the new identical graph to a blank area.
- Right-click on one of the graphs and choose **Scale>Density** from the context menu. The two graphs look very similar. Compare the graphs.

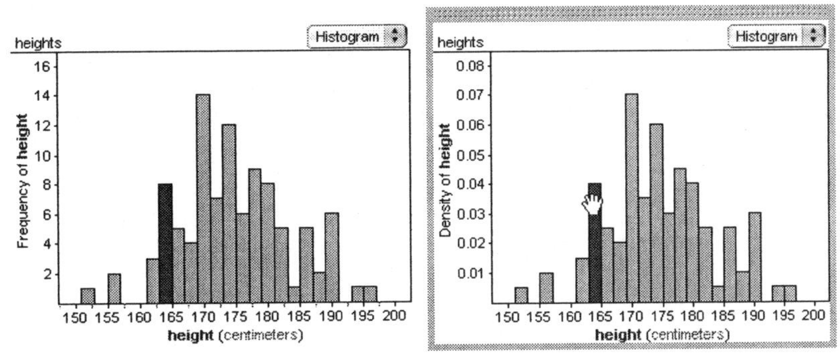

The difference is the vertical scale. In the illustration, the graph on the left plots *frequency*, which is the number of cases. On the right, we see *density*. This is now a *density histogram*, which has many of the same properties as a density curve. In particular, the area under the curve is 1.

We have selected a bar in the density graph, which automatically selects the same bar in the frequency graph. Its frequency is 8 and its density is 0.04. The 8 is easy to understand—eight people have heights between 163 and 165 centimeters[1]. But what does the 0.04 really mean? The key is to think about area. The area of the bar is 0.04 times 2, because 2 is the width of the bin. That product is 0.08. And that is the *proportion* of the 100 cases that are in the bin.

Now try this:

❖ In each graph, point at the boundary between two bins and drag to make the bins wider—about 5 units wide. Look what happens:

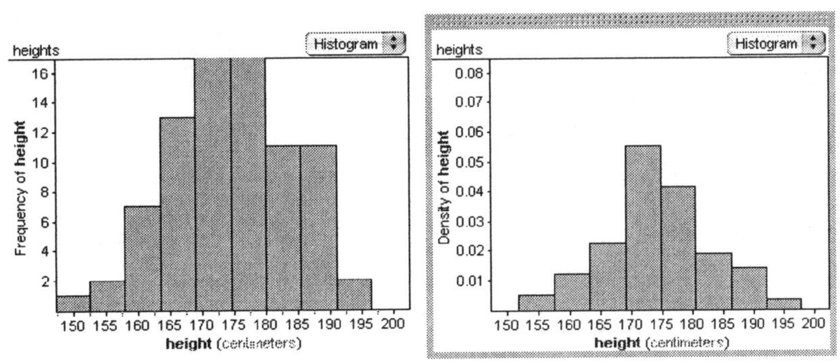

In the frequency histogram, the bars shoot out the top of the graph. When the bins get wider, the number of people in each bin increases. In the density histogram, the bars stay about the same height and smooth out. The area stays equal to one.

We see that density histograms have an interesting property: The appropriate scale stays the same no matter what you choose for the bin width. This is a good thing, and one of the reasons we would care about density at all.

1 As usual, including 163 but not including 165. Point to the bar and check out the status bar at the bottom of the screen.

One more thing: you don't see it on the graph, but the *units* of density in this case are "per centimeter." So when you calculate area, the units cancel and you get a pure, dimensionless number. That is, for the bar between 163 and 165 cm, the density of people is *4% per centimeter*. At that height, every centimeter range of height has about 4% of the population in it.

What about the curve?

Let's make the Normal curve and put it on our graph. Here are the instructions; you'll learn more of the details later on:

❖ Right-click the density histogram and choose **Plot Function** from the context menu. It's way down near the bottom.

❖ In the formula editor, enter **normalDensity(x, mean(height), stdDev(height))**. When you press **OK**, the function should appear, as in the illustration. The fact that it's called *normalDensity* is your clue that this is a genuine density curve, as described in the textbook.

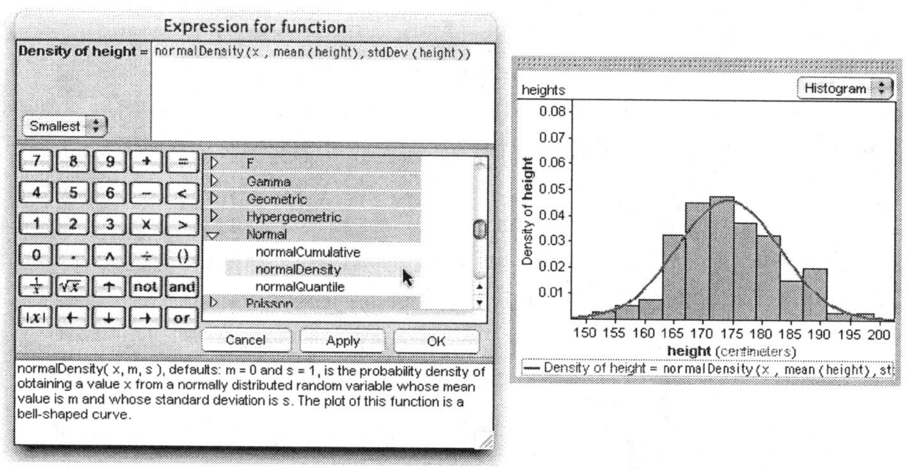

I CAN'T SEE MY WHOLE FORMULA! First, make sure the little menu in the upper left panel says **Smallest**. (It's the font size.) If that's not enough, drag the lower-right corner to resize the whole editor.

In the illustration, you can see the documentation for the **normalDensity( )** function in the *formula browser*, which is part of the formula editor. You'll find it under **Functions**, then under **Distributions**, then under **Normal**. Fathom has lots of functions, and each one has a description in the browser to help you.

Here are two more things to try:

❖ Drag the edges of the bars again to change the bin widths. Notice how the curve continues to approximate the data.

❖ Drag the **Sex** attribute to the *vertical* axis of the graph. The graph splits to show the separate distributions for males and females. In these data, the men are taller. Which sex has the bigger spread?

More exercises

2.14 Roll a distribution. Instead of using the calculator, we will make our simulation in Fathom. We'll cover simulations extensively in Chapter 6; think of this as a preview Note: In this setup, instead of rolling one die 120 times, we will roll 120 dice once.

❖ In a new Fathom document, make a table and give it one variable, **roll**.

❖ Right-click on the table and choose **New Cases...** from the context menu. Give it 120 cases.

❖ Now right-click on **roll** to give it a formula: **randomInteger(1, 6)**. (This means, pick an integer from 1 to 6 *inclusive*.) The table fills with random numbers.

❖ Make a histogram of the numbers. This is a frequency histogram.

❖ Duplicate the graph and turn it into a density histogram.

❖ Re-roll the dice by choosing **Rerandomize** from the **Collection** menu or by using its keyboard shortcut, control-**Y** (⌘-**Y** on the Mac).

(a) In theory, each number should come up 20 times. Why?

In practice, there is chance variation, so the histogram bars will have different heights. Theoretically, what should the frequency distribution look like?

(b) How is this distribution similar to the uniform distribution in Exercise 2.10? How is it different?

✐ How high are the bars in the density histogram? How high should they be theoretically? Explain.

❖ Right-click on the table again, and add 480 cases (for a total of 600).

✐ Now how high are the bars in the two graphs? What should they be theoretically? Why did the theoretical height change in one and not in the other?

2.21 Calculator-generated density curve. This is very much like 2.14, but we will use a different function to get a continuous set of random numbers.

❖ Follow the steps in 2.14 to make the table and new cases, except give the variable a more appropriate name and use the formula **random(0, 2)**. Make a density histogram of your results. Rerandomize a few times to see how they vary.

❖ Use your document to answer the questions in the textbook.

2.22 FLIP50. The file **02 Flip50.ftm** has the mechanisms for doing the simulation already built. When you press **Collect Measures**, Fathom flips 50 fair coins (in the **coins** collection), counts the number of heads, and ships that into the collection called **Measures from coins**. You can see the histogram being built as the results come in.

❖ Answer the questions in the textbook.

STOPPING A SIMULATION. You can stop Fathom from collecting new cases by pressing ESCAPE.

You will learn how to make files like this one in Chapter 8, but if you want to get a little early insight, open the collection inspectors for the various collections in the simulation files you see. There you can read (and edit) the formulas for the variables, and see what they do.

You can also look at different panels in the inspectors by pressing the tabs at the top. In **02 Flip50.ftm**, for example, inspect the **coins** collection and look at its **Measures** panel. You'll see a "measure" called **nHeads**; that's what the **Measures from coins** collection is collecting: one new measure—one value—for each set of 50 new coins.

To see how we made the nickels, check out the **Display** panel and its formulas.

2.2 Normal Distributions

In this section, we'll explore the Normal density curve and the Normal distribution.

Technology Toolbox: The Normal curve applet

Although you can use the applet on the Web to study the Normal curve, there is a Fathom file that you can use as well. The main thing Fathom does *not* do—in version 2, at least—is shade the areas under the curve. Other than that, this file will let you do more or less what you can do with the applet.

❖ Open the file **02 Normal Playground.ftm**. It will look like this:

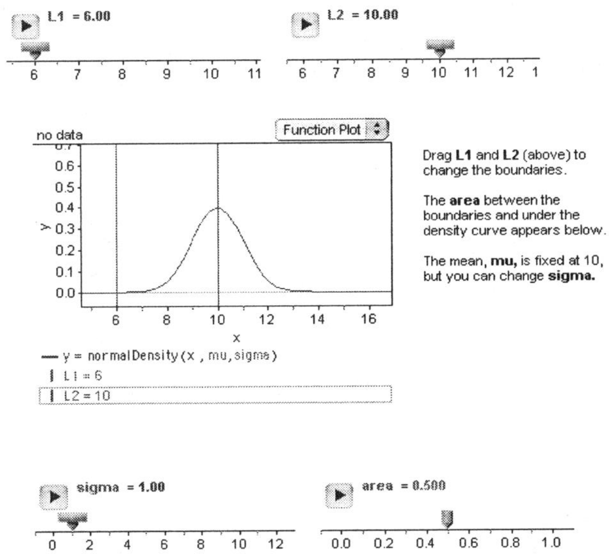

- ❖ You can move the sliders **L1** and **L2** to define a region under the Normal density curve, and the **area** slider will change to show how much area is in the region you have indicated. Use this to answer questions in the textbook (for example, question 1 on page 137 about checking the 68–95–99.7 rule).

Keep the following in mind as you answer the questions in the textbook:

- ❑ You can change the standard deviation by changing the **sigma** slider.
- ❑ You can change slider values by dragging the pointer or by typing. Dragging is dynamic. Typing is precise. To set **L2** at exactly the mean plus one standard deviation, type **11** into its value (assuming **sigma** is still at 1.0).
- ❑ We have fixed the mean **mu** at 10; there's really no need to change it in this context.
- ❑ Fathom's sliders come in two flavors. In some (like **L1**), you control slider values directly, while in others (like **area**), values are controlled by formulas. If the value is gray, it's computed—just like in tables. To see a slider's formula, inspect it by double-clicking on its axis or right-clicking and choosing **Inspect Slider** from the context menu.

Exercises

2.24–5 Men's Heights, extension. These two exercises give a "rule of thumb": The distribution of the heights of adult American men is approximately Normal with a mean of 69 inches and a standard deviation of 2.5 inches. Is that right?

- ❖ Open the file **02 Men 20-59 NHANES.ftm**. This contains height data on 174 American men from NHANES, the National Health And Nutrition Examination Survey, from 2003–04. (http://www.cdc.gov/nchs/nhanes.htm)

- Using a summary table, get one-variable statistics on the **height**s.
- Make a density histogram, and have Fathom plot a Normal density curve on it. Use a mean of 69 and a standard deviation of 2.5. (Use **Plot Function**, and **normalDensity(x, 69, 2.5)**).
- Then have Fathom plot an additional Normal density curve with the mean equal to the mean of the data set and the standard deviation equal to the standard deviation of the data set.
  - ✎ Do the data appear to be Normal?
  - ✎ Based on what you see, does the mean of 69 and a standard deviation of 2.5 fit? If you don't like those numbers, what would you use?
  - ✎ Describe at least two reasons why the rule of thumb in the textbook might be different from the actual data in the file.

2.27 FLIP50 MORE. Use data from the file **02 Flip50.ftm** to answer the questions in the textbook. As part of your "data analysis toolbox," you will need to add at least a summary table where you will have Fathom calculate statistics, especially the mean and standard deviation of **nHeads.**

Hint: Look in the **Measures from coins** collection for the variable.

Normal Distribution Calculations

The textbook has a box on pages 142–143 titled *Solving Problems Involving Normal Distributions* with four steps in it. You will use these steps many times as you solve problems. We have created a Fathom document to help. It's called **02 Std Normal Calculator.ftm**.

Let's use it on...

Example 2.9 Is cholesterol a problem for young boys?

- Open the file **02 Std Normal Calculator.ftm**.

First, we have to set up the right values for the problem. This can require some thought; be sure you understand what the problem is asking. In this case, we're looking at the distribution of cholesterol levels in boys. The problem states that cholesterol is Normally distributed, that the mean of the distribution is 170 mg/dL, and that the standard deviation is 30 mg/dL.

- Therefore set **mean** to **170** and **SD** to **30**. (It's easiest to do this by typing, not dragging.)

The problem goes on to explain that levels above 240 mg/dl may require medical attention. That's the place in the distribution we're interested in.

- Set **value** to **240**.
- The left-hand graph may now be blank; to re-scale the axes, choose **Scatter Plot** from the menu in the corner of the graph.

❖ Your file should look like this (after you have re-scaled your axes):

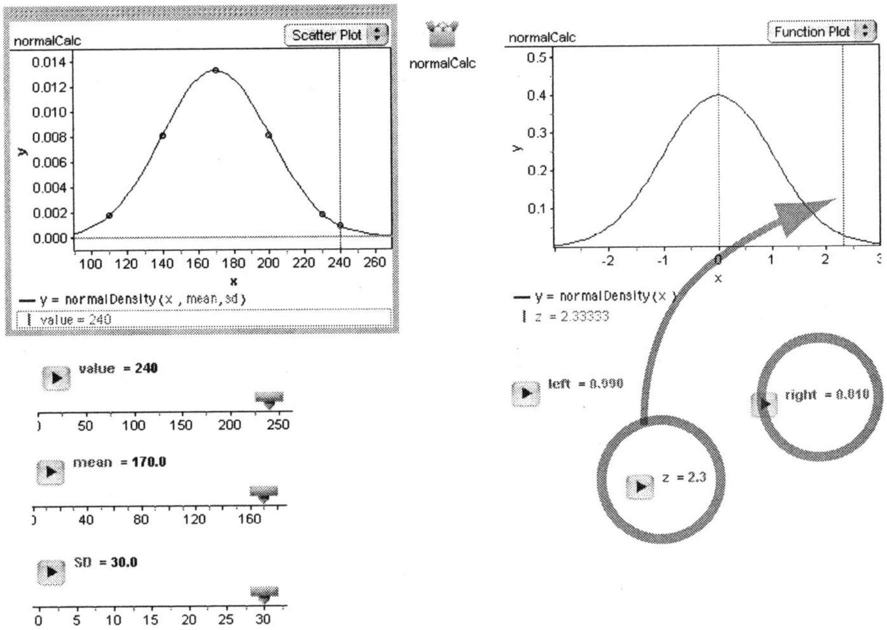

On the right-hand side, you can see the standard score, **z** = **2.3**, and its position on the standard Normal distribution. You can see that the **left**-hand area is 0.990, and that the **right**-hand area is 0.010, or 1%. If cholesterol really is Normally distributed as it is described, only 1% of the population has a dangerously high cholesterol level.

DON'T USE THIS FILE AS A CRUTCH. Even though it may be handy, be sure you can compute *z*-scores with your calculator, sketch Normal curves by hand, and use the tables to look up Normal proportions.

Example 2.11 SAT Verbal scores, inverse Normal calculations

You can use the same file for this problem. The question is, how high a score do you need to be in the top 10%? We're told that the mean is 505 and the standard deviation is 110.

❖ Enter the right quantities into **mean** and **SD** to set up the distribution.

❖ Set **value** to some plausible number, such as 600.

❖ Rescale the scatter plot so you can see the curve.

❖ Slide **value** until **right** (over on the right) is equal to 10% (i.e., 0.100).

The machinery of the file does the standardization and un-standardization for you, and, of course, all the table lookups. Again, don't let this be a crutch; but use it to improve your understanding of what's going on. Ask yourself key questions all the time, such as, "Why am I looking at **right**? What does **left** tell me? What does **z** mean in this context? What score do I have to get to be in the top 5%?"

Exercises

You can use the same file to help you with any of the exercises 2.31 to 2.35. Here's a little supplementary exercise to follow 2.34(c):

2.34 IQ TEST SCORES (EXTENSION).

(d) In 1979, *Omni* magazine published the "Langdon Adult Intelligence Test," which was designed by Kevin Langdon to detect even more exceptional people than the MENSA tests could. He was starting the "Four-Sigma Society," to be populated by people who (like him, naturally) had IQs four standard deviations or more above the mean.

✏ What IQ do you need on the Wechsler scale (described in exercise 2.34) to be in the 4-sigma society?

✏ In a city with 1 million adults, how many of them would you expect to qualify for this society?

❖ You will need to re-scale the **z** graph in order to see the 4-sigma line, and you will need to expand the **right** slider and re-scale it in order to get a number other than zero there.

Let's go backwards and look at Exercise 2.25:

2.25 STILL MORE ON MEN'S HEIGHTS. Remember how the textbook gave a rule of thumb that men's heights were roughly $N(69, 2.5)$? And how we looked at data in **02 Men 20-59 NHANES.ftm** and probably found that it was more like $N(69, 3.0)$?

Let's look again at question (a): What percent of men are taller than 74 inches? What you're supposed to say back there at exercise 2.25 is that it's 5 inches different, so that's 2 sigmas out, and there are 95% inside 2 sigmas, which means that only 2.5% are *above* 2 sigmas.

✏ If **SD** were 3 inches instead of 2.5, how many sigmas would you need to be 74 inches tall?

✏ What percent of people are taller than that?

✏ In the data file, what percent of the men have heights above 74 inches?

The point of this is that out in the tail of the Normal distribution, small changes in the mean or standard deviation can make a big difference in how many cases you expect to find. Just changing the standard deviation from 2.5 to 3.0—a change of only 17%—just about doubles the number of people you expect to find above 74 inches. You'll find similarly large changes if you change the mean a little.

Assessing Normality

You can use Fathom tools you're familiar with to see if the distribution is approximately bell-shaped. You can even make a density histogram and superimpose a Normal curve. There's another tool for assessing Normality, though: Fathom can make Normal probability plots; it calls them *Normal quantile plots.*

❖ Open **02 Men 20-59 NHANES.ftm**. The men's **height**s are approximately Normally distributed.

❖ Make a graph of **height** by putting height on the *vertical* axis.

❖ Choose **Normal Quantile Plot** from the menu in the corner of the graph. The graph changes from a dot plot to something new; see the illustration:

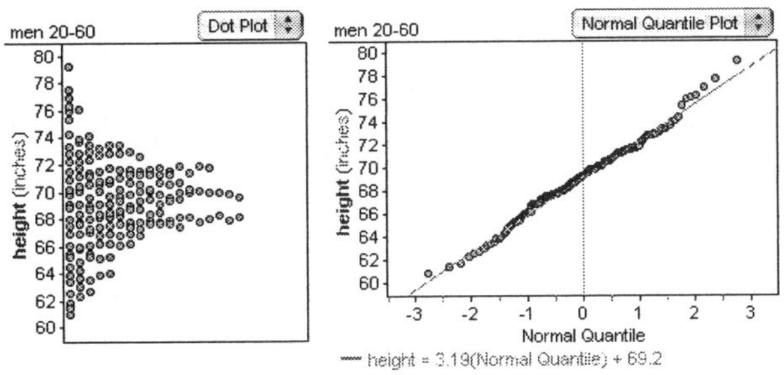

Basically, as described in the textbook, Fathom computes the z-score from the Normal distribution that corresponds to the percentile of every point, and plots it on the horizontal axis, against the variable we started with on the vertical axis, **height**. If the data are Normally distributed, this plot will give a straight line. Systematic differences indicate a non-Normal distribution. This is definitely close enough that we would call it Normal.

Fathom also fits a line to the points. Note that the slope of the line is the standard deviation (how much the variable changes if z changes by one) and the intercept is the mean. If you put the data on the horizontal axis (as you normally would) you can still get a Normal quantile plot, but the slope and intercept won't be as nice.

Example 2.13 Connecting wires and wafers

Use **EG02_13 Wires and wafers.ftm**. Look at the distribution, then make a Normal quantile plot and see how it matches the one in the textbook. See the comments there in the text.

Example 2.14 Guinea pig survival

You are to assess the Normality of the guinea-pig **survival** times data, which you will also find in the file **EG02_14 Guinea pigs.ftm**.

❖ Open the file and make a regular dot plot or histogram to look at the distribution. You should see *immediately* that the distribution is skewed.

❖ Make a Normal quantile plot; the illustration shows you what a skewed distribution looks like in a Normal quantile plot:

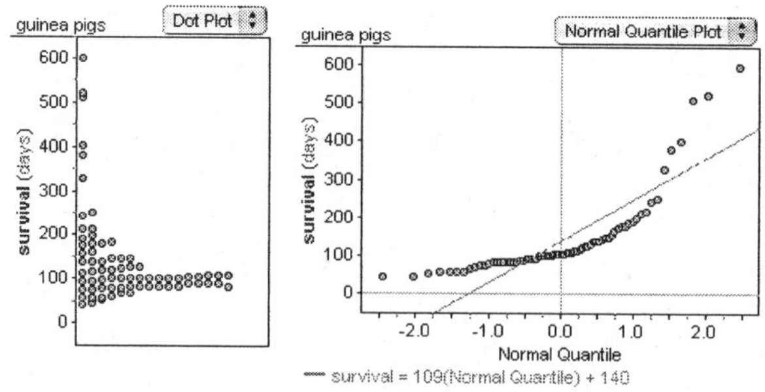

❖ Use the equation for the line to estimate the mean and standard deviation of the data.

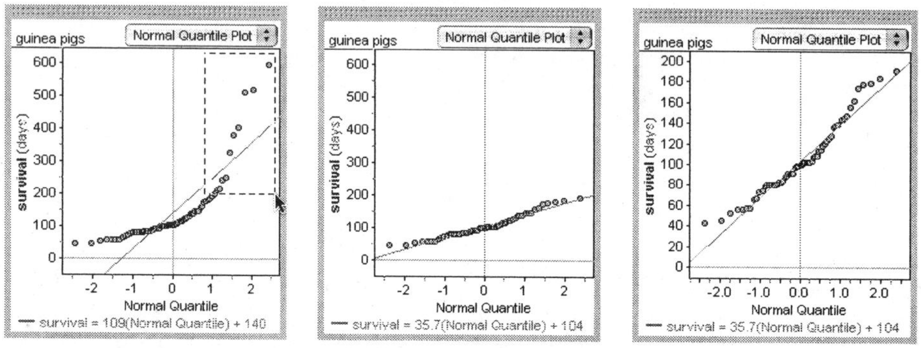

❖ It's the high-value points that make it not Normal; the others are much closer to being in a line. So select the top points in either graph (drag a rectangle around them) and **Cut** them (Use CONTROL-**X** or ⌘-**X**, not DELETE!) from the data set. The "fit" line will adjust. Rescale the plot to see how close to the line (close enough) the distribution now falls.

❖ Estimate the new mean and standard deviation.

❖ Be sure to paste the data back in—or close the file without saving, or **Undo** repeatedly—or you may lose those cut cases forever!

The illustration shows the original plot while we're selecting the top points; then the plot after the **Cut**; then the same plot, rescaled. Notice how the standard deviation decreased from 109 days to about 36.

Exercises

You can now use Fathom to help you with many of the exercises, especially 2.39 and 2.40, for which there are files already prepared. Let's look more closely, however, at…

2.41 NORMAL RANDOM NUMBERS. Do the tasks in the textbook, except use Fathom instead of your calculator to simulate the numbers:

- ❖ Make a new document and a table with one attribute, **x**. Give the table 100 cases.
- ❖ Give **x** the simple formula **randomNormal( )**. This fills **x** with numbers sampled from a standard Normal distribution, that is, with a mean of 0 and a standard deviation of 1.

❖ Graph the data (what kind(s) of graphs do you need?) and answer the questions. Use **Collection>Rerandomize** to get new sets of data.

2.42 UNIFORM RANDOM NUMBERS. Do the same as in 2.41, except use the simple formula **random( )**, which returns a uniformly distributed number between 0 and 1. Be sure to see how the Normal quantile plot for uniform data looks. Not Normal.

DEFAULT ARGUMENTS. Fathom's functions often have *default* arguments. For example, **random(5, 10)** returns a number between 5 and 10. The function has two *arguments*, 5 and 10. If you just use one argument, as in **random(10)**, that means between *zero* and 10. With no arguments—**random( )**—Fathom assumes you mean between zero and *one*. Even with no arguments, however, you still need the parentheses.

Different functions have different defaults. The function **randomNormal( )** means the same as **randomNormal(0, 1)**. Mean first, then standard deviation.

2.47 DECILES. Use **02 Std Normal Calculator.ftm**. Or, if you're adventurous, use random Normal numbers and an Ntigram! (Use at least 1000 points, and use the graph's inspector to set it to 10 bins.)

2.48 OUTLIERS IN A NORMAL DISTRIBUTION. A Fathom strategy: Make a collection with a whole lot of random Normal data values (see 2.41). Then make a box plot of them. See where the "outliers" start.

Fathom can help you with many of the remaining exercises. And while Fathom does not (yet) have an equivalent of the **shadeNorm** function on the TI, you can use **02 Std Normal Calculator.ftm** or **02 Normal Playground.ftm** to do much of the same work.

The Other Normal Functions

The textbook mentions using **ShadeNorm**, the Normal cumulative density function, and the inverse Normal function for the TI, in order to solve problems without looking things up in the table. How do you do that in Fathom?

The first answer is to use the files we have already talked about, **02 Std Normal Calculator.ftm** or **02 Normal Playground.ftm**. This assumes that you have your trusty computer handy and remember where those files are. But for some problems, the computer is more machinery than you need. So be sure you know how to use the tables, and be sure you know how to use the calculator. Given that, by all means use those Fathom files to help you solve problems and to visualize the Normal distribution as it comes into play.

If you look "under the hood" in those files, you can see how we use Fathom's cumulative and inverse Normal functions. Let's review the four interrelated Normal functions (of which we have only discussed two until now):

normalDensity(x, mu, sigma)	This returns the probability density of obtaining a value **x** from a Normally distributed random variable with mean **mu** and standard deviation **sigma**.
normalCumulative(x, mu, sigma)	This is the probability of drawing a number ≤ **x** from that same distribution.
normalQuantile(c, mu, sigma)	This is the value of a Normally distributed random variable whose cumulative probability is **c**. That is, this is the inverse of **normalCumulative( )**.
randomNormal(mu, sigma)	This returns a random number from the Normal distribution specified by **mu** and **sigma**.

Here are some of the ways you might use these functions:

Use **normalDensity( )** to plot a Normal density curve.

Use **normalCumulative( )** to find areas under a Normal curve. If you look in the Fathom files we mentioned, we use it a lot in the formulas, for example, the formula for the slider **right** is **1 – normalCumulative(z)**. (In all of these functions, **mu** defaults to zero and **sigma** defaults to one.)

Use **normalQuantile( )** to find the data value that corresponds to a particular probability or percentile. So if you want to know the 90th percentile of some exam scores that are distributed $N(100, 15)$, the function would be **normalQuantile(0.90, 100, 15)**.

Use **randomNormal( )** in simulations when you want to generate numbers from a Normal distribution.

3 Examining Relationships

Data instantly become more interesting when more than one variable is involved. The tools you learn in this chapter will help you describe and quantify relationships between two quantitative variables.

3.1 Scatterplots and Correlation

You already learned how to make scatterplots. You simply drag one quantitative variable to one axis, and one to the other. In general, you put the *explanatory* variable on the horizontal axis and the *response* variable on the vertical. Sometimes, as explained in the textbook, you're not sure which is which. In that case, in Fathom:

❖ Drop the two variables on the axes to make a scatterplot. Don't worry about which is which.

❖ If you want to switch them, just drag one variable name on top of the other. They swap.

Example 3.3 State SAT Math scores

Let's begin by reproducing Figure 3.1 in Fathom.

❖ Open the file **EG03_03 State SAT 2005.ftm**. (It's in the folder for Chapter 3.)

❖ Make a graph with **math_05** on the vertical axis and **pctTaking** on the horizontal.

That was straightforward; we can see the negative association described in the textbook. The way we assigned variables suggests that the average math score depends on the percent of students taking the test rather than the other way around. Despite that, remember that the graph only establishes the association, not causation.

One more thing: Remember that you can point at any data point with the mouse and get information about that point in the status bar. The illustration shows how Colorado has a higher average score (560) than others with a similar percentage of students taking the test.

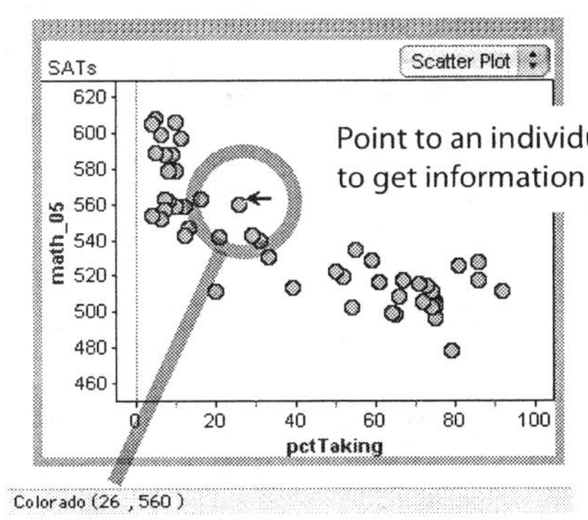

Example 3.5 Beers and blood alcohol

Now let's use the beer example. Student volunteers from Ohio State University drank a randomly assigned number of cans of beer; thirty minutes later, a police officer measured their blood alcohol content (BAC). Here are the data from the textbook.

Student:	1	2	3	4	5	6	7	8
Beers:	5	2	9	8	3	7	3	5
BAC:	0.10	0.03	0.19	0.12	0.04	0.095	0.07	0.06

Student:	9	10	11	12	13	14	15	16
Beers:	3	5	4	6	5	7	1	4
BAC:	0.02	0.05	0.07	0.10	0.085	0.09	0.01	0.05

❖ In a new Fathom document, enter the data. There are three variables: **Student**, **Beers**, and **BAC.** Enter them in *columns*, not rows, as in the table above.

❖ Make a new graph; drag **Beers** to the horizontal axis and **BAC** to the vertical axis.

Your graph should look like Figure 3.2 in the textbook. You should see a positive linear association.

Example 3.6 Is the South different?

We will use this example to learn about *legend attributes*.

Using the same file with SAT scores for the states (**EG03_03 State SAT 2005.ftm**), make a scatter plot of math scores as a function of percent taking. Then drop the **region** variable into the *middle* of the plot, as shown in the illustration. Fathom plots the different regions in different symbols, and inserts a *legend* at the bottom of the graph. What do you notice about the clumps of points?

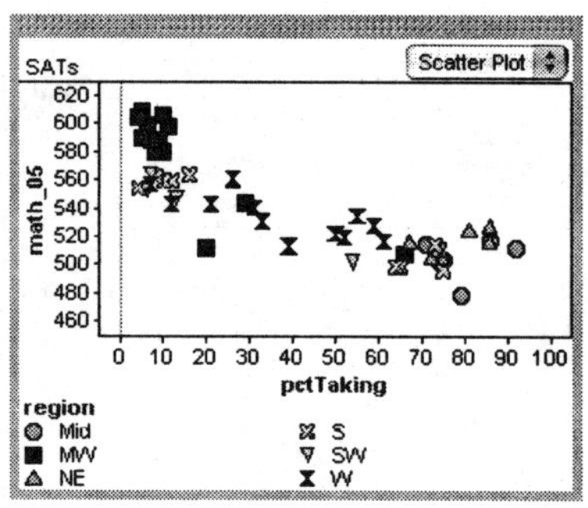

Click on one of the symbols in the legend at the bottom to select all of those points. Use this feature to select different regions to see if any region does worse than others.

Now drop a continuous attribute (such as **verbal_05**) into the middle. Fathom makes a "rainbow" scale in the legend area. You can see by the colors that verbal scores more or less follow the math scores.

Finally, to get rid of the legend, right-click to get the context menu and choose **Remove Legend Attribute....**

Exercise 3.5 The endangered manatee I

Doing the problem doesn't require special Fathom knowledge.

❖ Enter the data from the textbook or use the file **EX03_05 Manatees.ftm**.

❖ Answer the questions in the textbook.

While we're here, however, we can learn a couple of new things.

❖ You probably have the number of boats (**registrations**) on the horizontal axis and the number of manatees **killed** on the vertical. Try switching the two; drag one axis label onto the other, then put it back.

❖ Try putting **Year** on the horizontal axis. Now you see how the number of manatees killed changed by year. The graph should look similar to the previous one.

More about Manatees: Connecting the Dots

Beware! We're about to discuss something with which your instructor may legitimately disagree. In case of inconsistency, follow your instructor.

Here's the issue: When is it OK to connect the dots? We will soon be putting lines on graphs like these—straight lines that approximate the pattern of the points but do not actually go through them. Some students feel the urge to connect the dots in all sorts of graphs, and this drives teachers crazy.

In a case like this, however, many people (such as the author) feel that it is OK to connect the dots with line segments. Why? In a time series, the line segment itself shows something relevant: the trend from one year to the next. Try it.

❖ Choose **Line Scatter Plot** from the menu in the corner of the graph.

The graph shows how the number of kills increases with time; each line segment shows the annual trend.

❖ Now make the graph show kills as a function of registrations again. The graph is no longer a time series.

In this case, we all agree, the lines are bad[1]. In general, resist the temptation to play dot-to-dot with your scatter plots. Time series are often an exception.

1 But not always. The lines still connect one year to the next, so we'd be justified if we were trying to show how the relationship changed over time. But that's more complicated than our situation—we're really just interested in the relationship between registrations and kills.

❖ Now drop the number of **registrations** on top of the number of kills, replacing the variable. You should see the pattern of boat registrations over time, a very smooth graph.

When we have **Year** (or any time variable) on the horizontal axis, our graph is called a *time series*. This is a scatter plot, to be sure, but it shows how things have changed with time rather than simply showing an association.

Here are two things to think about now that you have looked at the time series:

✎ First, how would you describe the time series showing the registrations? How is it qualitatively different from the time series of kills? How would you explain it?

✎ Second, how do we know that the deaths of the manatees are due to increased registrations? Maybe manatees are just naturally dying faster.

More Exercises

Use Fathom to work many of these exercises; enter the data yourself from the textbook, or use files from the CD with appropriate names. Exercise 3.42, for example, will begin with **EX03_42**.

3.10 Do HEAVIER PEOPLE BURN MORE ENERGY? To get just the female subjects, use a *filter* on the graph. We learned about filters in Chapter 1; see page 14. A good expression for the filter might be **Sex = "F"**. Don't forget the quotes!

Calculating Correlation

In Chapter 1 you learned how to make calculated columns. Now we're going to use that skill, coupled with summary tables, to calculate the correlation coefficient as described in the textbook. Then we'll learn the shortcut.

❖ Enter the backpack data from the textbook into Fathom columns. As usual, enter them in columns, not rows! "Weight" is the student's weight. If you want to use units, enter them with the data, e.g., **120 lbs**.

Weight (lbs)	120	187	109	103	131	165	158	116
Pack (lbs)	26	30	26	24	29	35	31	28

❖ Make a scatter plot (always graph your data if you can!) and see the positive association. Put **weight** on the horizontal axis and **pack** on the vertical.

❖ Make a new column, **zWeight**, which will be the *z*-score of **weight**. As a reminder, to do that:

  ❖ Right-click on the name **zWeight** and choose **Edit Formula** from the menu.

  ❖ Enter **(weight – mean(weight)) / stdDev(weight)** as the formula.

  ❖ Click **OK** to close the editor. The variable **zWeight** should fill with reasonable values.

❖ Make another new variable, **zPack**, and give it an appropriate formula.

❖ Make yet another new column, **prod**. Make it the product of **zPack** and **zWeight**.

❖ Drag **prod** to a new summary table. The mean will appear.

❖ From the context menu, choose **Add Formula** and enter **sum( )/(count( ) – 1)**. That's *r*, the correlation coefficient: the sum of the products of the *z*-scores, divided by one less than the number of cases.

❖ **Save** your file! You'll need it in a few pages.

Once you have the data and the table set up, you can explore:

❖ Make a dot plot of **prod** and see which points in the **prod** plot correspond to which points in the scatter plot. Be sure you understand the correspondence.

❖ Drag the points in the scatter plot and see how they affect the **prod** plot and the value of *r*.

❖ Try really hard to make *r* greater than 1 or less than –1.

Extra! Extra! Making a computed slider

You can see in the summary table how the correlation coefficient—the sum of the **prod**s—changes as you drag points, but it would be good to see that visually. We're going to make a slider that will act like a meter, always showing you the current value of *r*.

❖ Drag a new slider off the shelf. Change its name to **r**.

❖ Drag the *name* of your collection onto the slider and drop it. *This is a crucial step!* You're linking with the collection, giving the slider access to the data.

❖ Double-click the slider to open its inspector.

❖ Double-click the formula box to give the slider a formula.

❖ Enter **sum(prod)/(count( ) – 1)**. Click **OK** to close and see how it works.

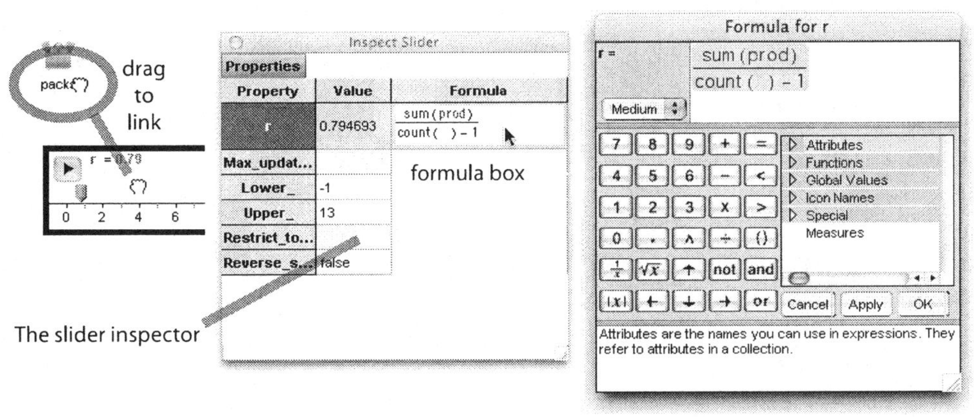

Shortcuts

The *z*-score is so important that Fathom has a function that does it for you: **zScore( )**. That is, you could use **zScore(weight)** as the entire formula for **zWeight**. The correlation coefficient is also important enough to have a function of its own:

❖ Make a new summary table.

❖ Drag **weight** to the left side and **pack** to the top. Shazam! The correlation coefficient! The function is called **correlation( )**. If you're not in a summary table, you must specify both variables as arguments, as in **correlation(weight, pack)**.

Activity 3B Correlation and Regression Applet

Follow the instructions in the textbook, but use the file **A03_B CorrelationPlayground.ftm**. It has two sliders: The **n** slider controls how many points you can see; the **r** slider reports the current value of the correlation coefficient.

3.2 Least-Squares Regression

Regression is a process for finding a model given a set of data. When the model is a line, we call it a *regression line*. When we use the process called *least squares*, we get a least-squares regression line, often abbreviated *LSRL*.

Example 3.10 Fat gain and NEA

We will use this example to see how to make an LSRL in Fathom.

❖ Enter the data from the textbook. (You can also see it in the illustration that's coming.)

❖ Make a scatter plot showing the relationship between the variables. (We called them **dNEA** and **FatGain**. Call them what you like; just make the names sensible.)

Fidgeting and weight gain

	dNEA	FatGain	<ne
1	-94	4.2	
2	-57	3.0	
3	-29	3.7	
4	135	2.7	
5	143	3.2	
6	151	3.6	
7	245	2.4	
8	355	1.3	
9	392	3.8	
10	473	1.7	
11	486	1.6	
12	535	2.2	
13	571	1.0	
14	580	0.4	
15	620	2.3	
16	690	1.1	

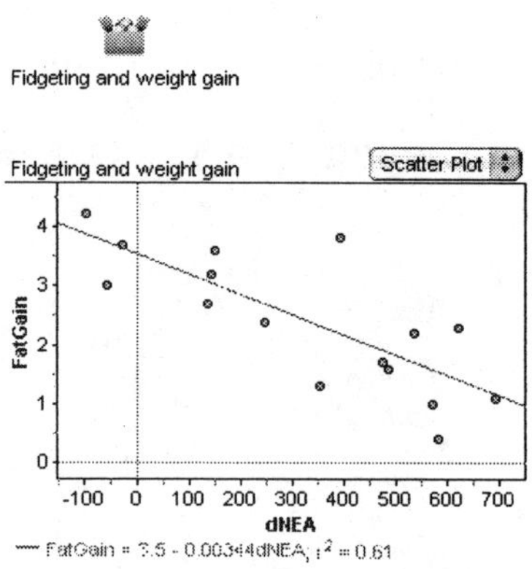

❖ Right-click on the graph to get the context menu and choose **Least-Squares Line**. The LSRL appears on the graph, and the formula for the line appears below it. The text there also shows the *square* of the correlation coefficient, r^2. In this case, its value is 0.61.

If your formula is in $y = mx + b$ form, you can change it to the form you use in the textbook.

❖ Choose **Preferences...** from the **Fathom** menu. There, adjust the **Linear Equation Form** to **y = a + bx**. (See the illustration.)

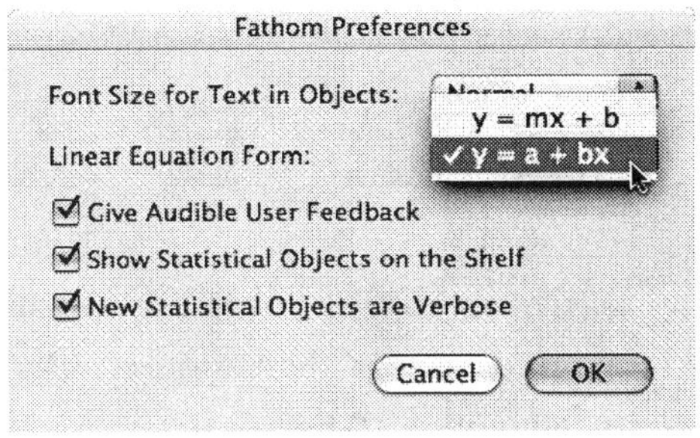

Knowing this, you can do many of the examples and exercises in this section. If the files are on the CD, their names will follow the convention we described earlier.

But here is a trick that will help. The formula above is

FatGain = 3.5 – 0.00344 dNEA.

Figuring out what the numbers mean can be a problem. But if you put *units* into your data (right-click on the table; choose **Show Units**; type **cal** and **kg** into the "units row") the meaning may become more intelligible. The equation for the line will change to:

FatGain = 3.5 kg + (–0.00344 kg/cal) dNEA.

Ah! If you can get past the parentheses and the fact that units are there at all, you can see that the 3.5 is in kilograms. It's what you get when the change in fidgeting (**dNEA** in our example) is zero. What could it mean? It's the no-change weight gain—three-and-a-half kilos. The slope is even easier. If you say, "minus .00344 kilograms per calorie," the meaning almost explains itself: on average, you lose 3 grams for every additional calorie you burn.

The Meaning of Least Squares

Let's see why they call it "least squares."

❖ Open up the file you made with the backpack data. (You did that on pages 50–51).

❖ Make a scatterplot where pack weight is the response variable.

❖ *Don't add a least-squares line!* Not yet, anyway. Instead, use the context menu for the graph to add a **Movable Line**.

❖ Drag the line to "eyeball" a fit to the data. This is just like using a clear ruler on a graph you have made by hand. Notice how the equation for the line changes as you drag.

❖ Now choose **Show Squares** from the context menu. Squares appear connecting the line to the points. Also, a number appears below the graph—with the equation—labeled "Sum of squares."

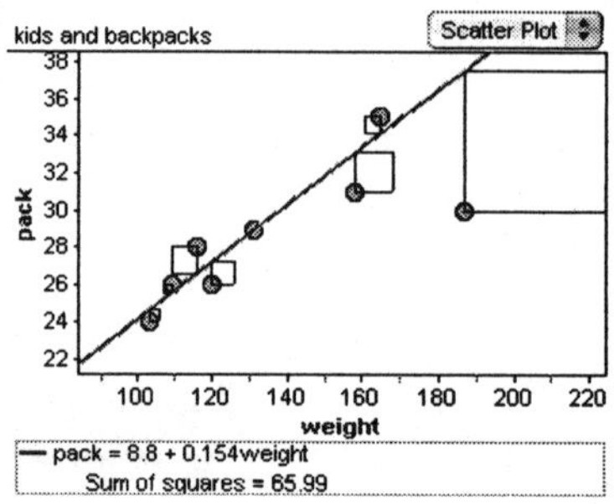

This number is the sum of the *areas* of the squares. You want the line to go close to the points, but how can you tell if one line is better than another? One way to decide is to pick the line where the squares are smaller, and have less area.

❖ Therefore, move the line to make this "Sum of squares" number as small as possible. You may have to alternate several times between changing the slope (by dragging the ends of the line) and changing the intercept (by dragging the middle).

❖ Now add the least-squares line. It has the minimum area possible. (It's literally the line with the *least* area in the squares.) See how close your movable line came; if it was not close, figure out why, and practice with other data.

SQUARE OF WHAT? In least squares, the thing that gets squared is the *vertical* distance from the line to each point. This distance is called the *residual*. It's what's left over in the data after you take out the model (the line).

Playing with the Points

You can play with the points, the squares, and the lines in order to better understand how they interact (this supplements Activity 3C with the Correlation and Regression Applet).

❖ Remove the movable line by using the context menu. If the squares get confusing, remove them, too.

❖ Drag the point that's farthest to the right more in line with the other points. What happens to the line? To the value of r^2? To the sum of squares?

❖ Drag the same point down and to the right. Try to make the slope go negative. You may need to change the scale of the plot so you can drag it farther.

❖ When you have the equation really out of whack due to that one point, select the point and **Cut** it. The line will snap back to fit the remaining points.

❖ Keep experimenting.

❖ When you're done, use **Undo** repeatedly (learn its keyboard shortcut—CONTROL-**Z** or **⌘-Z**) until the data points are all back where they were originally.

THE LEAST-SQUARES LINE IS EASILY AFFECTED BY OUTLIERS. A single outlier can completely change the line you calculate, even changing the direction of the association. The lesson? *Always plot your points.* Don't just calculate blindly!

The Means

We want to see how this line relates to the means of the x and y variables. In our case, these are **weight** and **pack**, respectively.

❖ Continue with the data on kids and their backpacks. If the file is too messed up, close it *without* saving, open a new copy of the file, make the graph, and add a least-squares line.

❖ From the graph's context menu, choose **Plot Value** (not **Plot Function**).

❖ Enter **mean(weight)** into the equation editor and press **OK**. This plots a vertical line at the value of the mean of the x-variable. If you don't use **weight**, change the formula, of course.

❖ From the context menu again, choose **Plot Function**. This time enter **mean(pack)** (or whatever you used). This plots a *constant* function equal to the mean of the y-variable.

❖ Now drag points around and see what you notice. Again, **Undo** when you're done to restore the data.

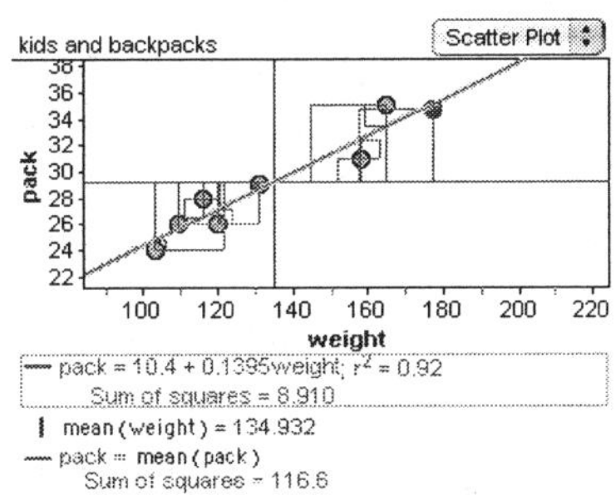

Making Predictions

Sometimes you will want to see what y-value the LSRL predicts for a given x-value. There are a number of ways to do this in Fathom. This one is crude, but effective:

❖ Make a new column (a new variable) named **pack_hat**. We use "hat" because we often put a hat over a variable name to indicate that it is a predicted value, something from a model.[1]

❖ Give **pack_hat** a formula that's identical to the equation for the least-squares line. Just type it in.

---

1 As you will see, we also use it when it's a value from a *sample*, which is an entirely different thing. This kind of inconsistency can drive you nuts. Get used to it.

Now the values in **pack_hat** are the "predicted" weights of the packs for the kids. To see what this means, do this:

❖ Drag **pack_hat** to the same axis as **pack**, dropping it on the plus sign that appears. Now they're both plotted on the same graph, and you can see that the **pack_hat** points are on the line.

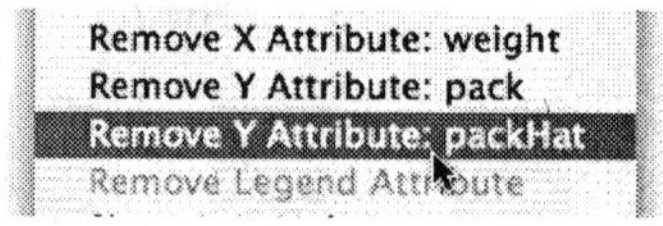

❖ Get rid of that attribute using the trusty context (i.e., right-click) menu.

You can also use the Fathom data table to make specific predictions. For example, suppose a kid weighing 150 pounds wanted to know about how much his pack should weigh.

❖ Enter 150 under **weight** in a new row (a new case).

The variable **pack_hat** fills in with a "predicted" weight, even though there is no value for **pack**. No new point appears, because **pack** is still missing.

Alternatives to Least Squares

Although the least-squares line is useful and justifiably popular, it's just one way to make a regression line. There are other ways. For example, you could use the *median-median* line. You will find that line in the context menu of your scatter plot; make one and see what it looks like.

We won't study the median-median line here, but try messing with the data and see how it affects this "alternative" line. Like the median in contrast to the mean, the median-median line is more resistant to outliers than the least-squares.

Exercises

As usual, you can now use Fathom to help you with many exercises. Even so, be sure that you calculate a least-squares line by hand, at least once, to see how it works. In practice, you will use your calculator or a computer almost every time.

3.33 The Endangered Manatee II. You can use the file **EX03_33 Manatees 2.ftm**. The data that you are to add is in a text box in the file.

Residuals and Residual Plots

In the textbook, we're on page 214. You now know enough to use Fathom to work with residuals as described in the textbook. The main steps are:

❖ Calculate model values (e.g., **pack_hat**). These go into a new column with a formula (from the least-squares line, perhaps) that depends only on the explanatory (x) variable (e.g., **weight**).

❖ Calculate residual values. These go in another column, named something like **resid**, that has a formula to calculate how far the observation is from the prediction. You would use, for example, **resid = pack – pack_hat**). Be sure to get the sign right: If the point is above the line, the residual is positive.

Once you have the residuals, you can do all sorts of things—such as making a residual plot (**resid** against **weight**, say), and verifying that the mean residual is zero.

Residual plots are built into Fathom so you don't have to make separate columns for models and residuals and write all those formulas. The command is (where else?) in the context menu of the graph. Let's see how it works:

Example 3.14 Fat gain and NEA, again

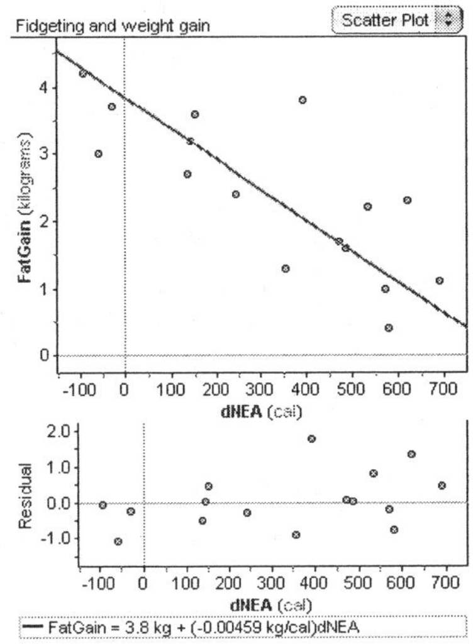

- ❖ Open the file **EG03_09 Fidgeting.ftm**. Plot **FatGain** against **dNEA**.
- ❖ Put a **Movable Line** on the graph. Make it roughly fit.
- ❖ Choose **Make Residual Plot** from the context menu of the graph. A residual plot appears at the bottom of the graph.
- ❖ Because things are a little squashed, stretch the graph vertically. Your plot should look roughly like the illustration.
- ❖ Now drag the movable line. You will see the residuals change. *Be sure you understand why the residuals move the directions they do.*
- ❖ Adjust the movable line, looking at the residuals, so that there is no particular pattern in the residuals and so that they are balanced above and below the line. (In the illustration, you can see an upward slope—a positive association—in the residual plot.)
- ❖ When you have it adjusted as well as you can, add the least-squares line. It probably matches pretty well. Most people come pretty close to the least-squares line by eye when they use residuals.

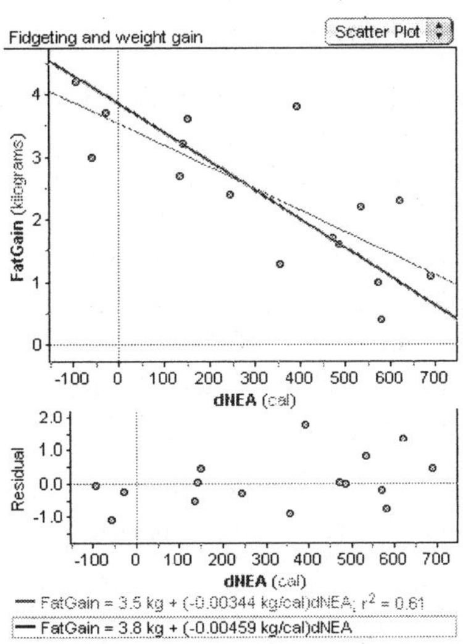

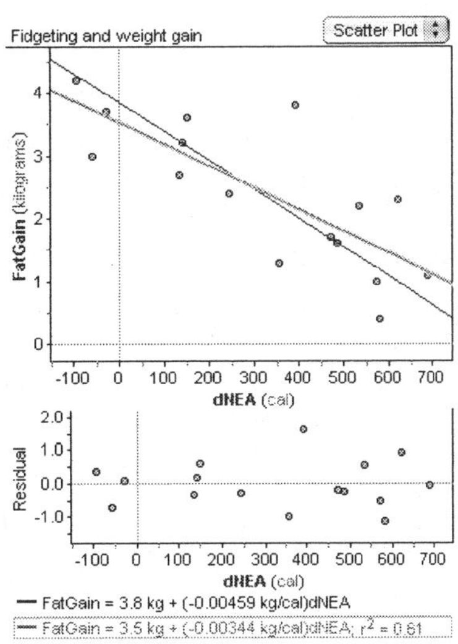

In the illustration, the residuals on the left are from the movable line, and they're tilted. The ones on the right are from the LSRL, and they're flat. Selecting the formula below the graph also makes the line thicker in the graph and puts a box around the formula.

Residual Plot Principle

The text makes important points about residuals that bear repeating (and amplifying). When a function does a good job of modeling data, its residual plot has three attributes:

- ❑ It shows no systematic patterns.
- ❑ It is centered (vertically) at zero.
- ❑ The residuals are relatively small.

Residual Plot mechanics

A residual plot in Fathom is just like any other graph, but it's stuck to its parent. Some things to note:

- ❖ You can get rid of a residual plot using the context menu.
- ❖ If you can't see the points, re-choose **Scatter Plot** from the popup in the corner of the graph; the plots will all re-scale to show all of the points.
- ❖ Rescale the vertical axis of the residual plot as you would any other axis.
- ❖ If you have more than one function (e.g., a movable line and an LSRL), you'll need to tell Fathom which one you want the residuals from. To do this, select one of the formulas by clicking (once) where it appears at the bottom of the graph. Select the other formula to compare.

Challenge: Example 3.15 Hubble's expanding universe

Example 3.15 in the textbook shows how least-squares regression is asymmetrical: The line when you try to predict *y* from *x* is different from the one when you try to predict *x* from *y*. The textbook shows the two lines. How did they get them? You try it:

- ❖ Open **EG03_15 Hubble data.ftm.** Make two scatter plots: **r** versus **v** and **v** versus **r**.

THE ORDER OF VARIABLES. When you say, "plot *a* versus *b*," that means that *a* goes on the *vertical* axis. Put the response variable *first* in a "versus" statement.

- ❖ Put a least-squares line on each graph.
- ❖ Figure out the formula that one line would have if plotted on the other graph. This requires algebra!
- ❖ Plot each function on its "opposite" graph. (Use **Plot Function** in the context menu.)

Your result should confirm Figure 3.22 in the textbook. The point is that if you switch the axes, the least-squares line changes.

Exercises

Again, you can use Fathom to help you with a number of the exercises that complete Section 3.2. One note:

3.43 REGRESSION IN THE LAB. If you enter the data, enter each observation as a new case—a new row in the table. This means that you will repeat each value of **Amount** five times.

3.3 Correlation and Regression Wisdom

This section elaborates on what you have learned about correlation and regression. Using Fathom can help illuminate these issues; we'll point out a few specific things.

Examples 3.16 and 3.17 Gesell scores

You will find the data in **EG03_16 Gesell.ftm**.

❖ Make the scatterplot of Gesell **score** versus **Age** (at first word).

Note that Fathom does *not* indicate that two of the points are "hiding" because of identical values.

Double-click an outlier. This opens the *inspector* to that case. You can see that, indeed, children 18 and 19 are the two outliers. In the illustration, you can see we have just double-clicked the outlier in *x*, and you can see in the inspector that this is Child 18.

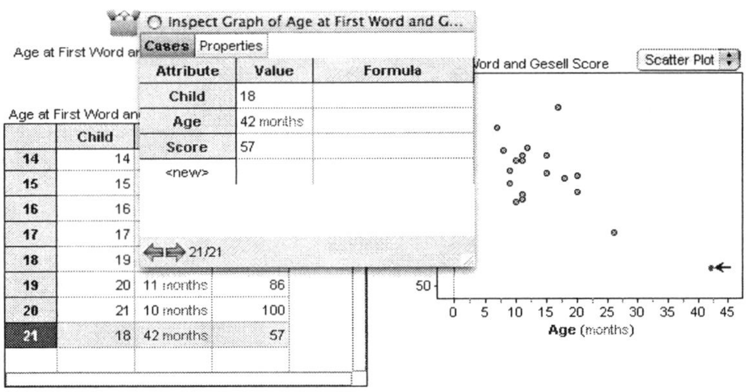

❖ Make a least-squares regression line. See that $r^2 = 0.41$, as advertised.

❖ Click once on Child 18 to select it. The point turns red.

❖ Cut the point (**Edit>Cut**, or CONTROL-**X** or ⌘-**X**). It disappears. What happens to the line? What happens to r^2?

- Paste the point back in (**Edit>Paste**, CONTROL-**V**, or ⌘-**V**) to restore the data. You could also use **Undo** (CONTROL-**Z** or ⌘-**Z**).
- Do the same with Child 19, the vertical outlier. See how much r^2 changes.
- Figure out how to cut them *both* out and restore them to see what happens to r^2. (Two strategies: Select both by SHIFT-clicking; or **Cut** them one at a time and use multiple **Undo** to restore them.)

Our statistical purpose is to explore what it means for a point to be *influential*: Child 18 influences the regression a lot, while Child 19 does not. Without Child 18, the model does not account for much of the variation in Gesell scores.

In Fathom, you can use **Cut** and **Paste** to explore what happens if data appear and disappear. Just be sure to put the data back!

4 More about Relationships Between Two Variables

In the last chapter we developed tools for describing the relationship between two variables graphically and with numbers. But we limited ourselves to quantitative variables that were related linearly. In this chapter, we move beyond that in several different directions.

4.1 Transforming to Achieve Linearity

There are several ways to explore the relationship between two quantitative variables whose relationship is not linear. We will follow the textbook's lead: we will transform the data so that we see a linear relationship, and then use our linear toolbox (e.g., the least-squares line) to nail down the function that relates the variables.

The basic strategy for this in Fathom is:

1. Plot the data and decide what transformation makes sense.
2. Make a new variable (or variables) for the transformed data.
3. Give each new variable an appropriate formula to transform it.
4. Plot the transformed variables and use a least-squares line to determine the linear function relating them. If the linear fit doesn't look good, try a different transformation.
5. Back-transform the line to get the nonlinear function in terms of the original variables.

You will be able to use Fathom to help you on many of the exercises in this section. We'll show you an example you can step through, then let you loose. This is something Fathom is *really* good at.

Then, later, we'll show you an alternative to linearizing.

Example 4.2 Fishing tournament

We'll use the fishing tournament from the textbook as our example of "linearizing" data.

We're trying to determine the weight of some rockfish without weighing them. We can use the length of the fish to find the weight. Our reasoning goes like this: We'll assume that all fish are the same shape even if they are different sizes. So if a fish is twice as long, it will also be twice as wide (up and down) and twice as thick (side to side). That means it will have *eight times* the volume. In general, the volume of the fish will be roughly proportional to the *cube* of the length. If all fish have the same density (which had better be close to the density of sea water!), their weight will also be proportional to the cube of their length, or $W = kL^3$.

Here is the key thing: This means that if we plot W against L^3, we should get a straight line (with slope k).

How do we do that in Fathom?

❖ Open **EG04_02 Rockfish.ftm**. Plot **Weight** vs. **Length**. (Remember: That means **Weight** is on the vertical axis; it's the response variable.) Note how the points curve up.

❖ We need the cube of the length. Make a new variable, **LengthCubed**. Give it the formula **Length^3**.

EXPONENTS IN THE FORMULA EDITOR. You can use the little caret key on the keypad [^] or press SHIFT-6 (on U.S. keyboards) to move into "exponent land." To get out of exponent land, type a right arrow, or click where you want to type.

❖ Now make a new graph of **Weight** as a function of **LengthCubed**. You should see a roughly straight relationship, as shown in the illustration. Remember that if you have both graphs visible, you can select points in one and see which ones they are in the other.

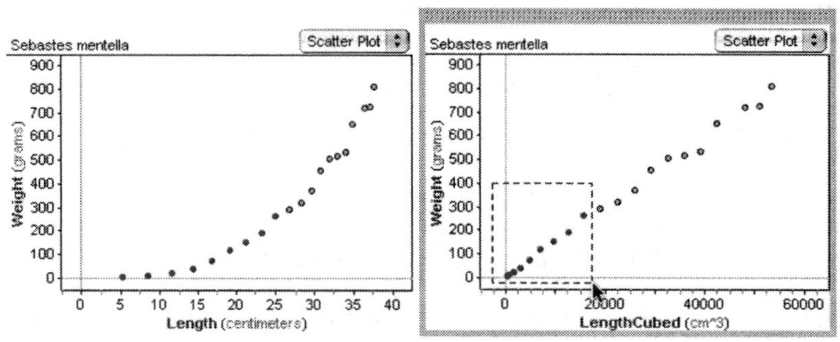

❖ Now put a least-squares line on the right-hand graph. Give it a residual plot.

❖ Notice how the intercept is not zero. If we're serious about the line going though zero, we can choose **Lock Intercept at Zero** from the context menu. The line will now go through zero exactly, and the formula will reflect this.

The next illustration shows the graph with residuals, with the "normal" intercept and with it locked. Locking the intercept doesn't make much difference. Either way, the formula is very close to **Weight = (0.015) * LengthCubed**. The residual plots don't show a systematic pattern except that the residuals are smaller for smaller fish, and the slope for the first few points looks negative in the residual plot, which means that the slope is smaller there in the data.

Now we're going to *back-transform* the linear equation to get its nonlinear counterpart. This is not as strange as it sounds. The strategy is to write the formula for the line, and substitute in the expressions we used to transform the data in the first place. In this case, it's easy: since **LengthCubed = Length^3**, that means $W = 0.015L^3$, that is, $k = 0.015$.

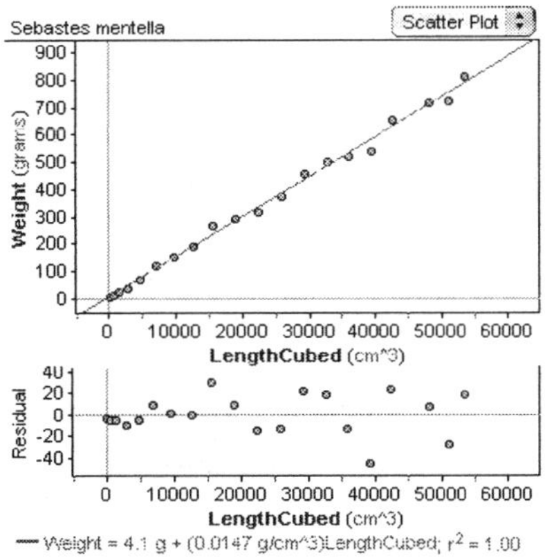

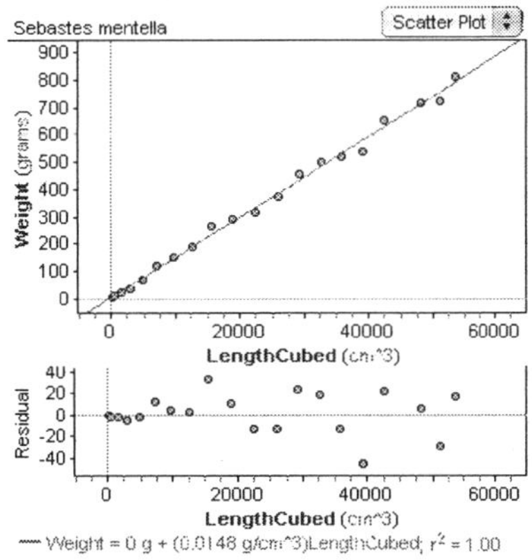

So (and this is the crucial part) we can plot the curved function on the original, untransformed graph by plotting **(0.015) * Length^3**.

❖ Context-click on the *original* **Weight-Length** graph and choose **Plot Function** from the bottom of the menu.

❖ Enter **0.015 * Length^3** as shown and press **OK**. You'll see the curve appear, going through the points pretty well.

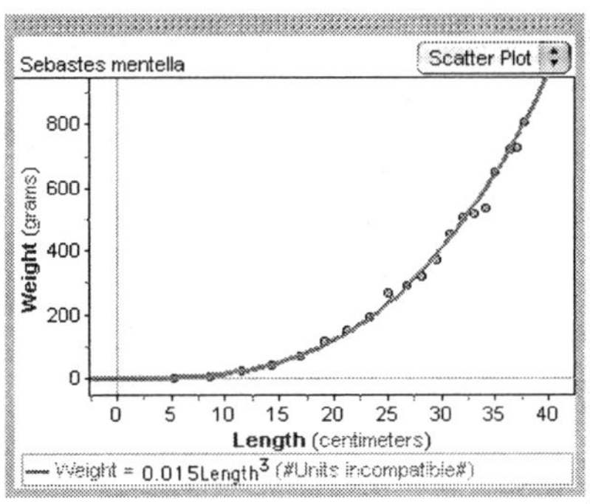

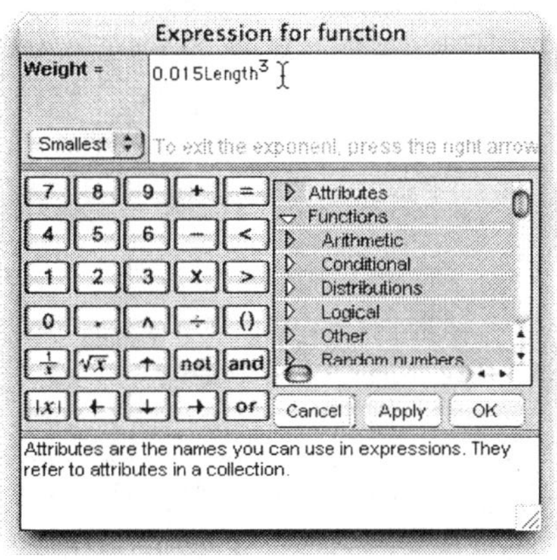

How did that work? We guessed the form of the function ahead of time; all we needed was the value of a *parameter*. We got that value, 0.015, from the least-squares regression in our transformed graph.

More things to do

❖ The **units incompatible** error is because we have **Weight** on one side, which is in grams, and **Length** cubed on the other, which is in cubic centimeters. Put the right units on the 0.015 (i.e., **0.015 g/cm^3**), and that message goes away.

❖ Make a residual plot for the curve as well. How is it different from the one we made for the line?

Some questions:

✎ Why should the intercept be zero?

✎ The textbook did not use a zero intercept. Why not, do you suppose? How much difference does it make? (You can give specific examples to help.)

✎ The parameter *k*, with units, is about 0.015 g/cm^3. Those are units of density. But fish ought to have a density close to that of water, which is 1.00 g/cm^3 (actually about 1.03 g/cm^3 for sea water). Two questions: Why is a value as small as 0.015 plausible? How do we know that the density of a fish is about the same as the water?

Exercises

For any of the exercises, be sure to ask yourself if the line though the transformed data should go though (0, 0). Here are some hints and other interesting tasks:

4.1 GONE FISHIN'. You will find a good linear relationship for $\sqrt[3]{W} = pL$ for some value of *p*, the parameter. When you back-transform the data to plot it on a graph with **Weight** on the vertical axis, you have to solve for *W before* you enter the formula. This changes the coefficient *p*. You should get a result close to what we got before.

4.1A SNOOK! There is nothing special about *Sebastes mentella*. Open the file **EX04_01A Snook.ftm**. This contains a lot of data about individual fish, in this case, snook. Find a cubic model that predicts **Weight** from **Length**. Find the value for the parameter *k*. And then answer this: What do the different parameter values tell you about the difference between snook and rockfish?

4.2 THE SWINGING PENDULUM, I. Extension: pretend you do not know the acceleration of gravity. Find a value for that parameter that best fits the data. By the way, Fathom understands the constant **pi**.

4.3 BOYLE'S LAW. Be sure to make two graphs, the original one and the transformed one, and then select some points at the extreme end of one of the graphs. Which points get selected in the other graph? Why?

4.5 LIGHT THOUGH THE WATER. We've now discussed exponential functions and logs, so let's take our time with this one and give you some more background in Fathom. First of all, Fathom's log functions are **log( )** for a common logarithm (base 10) and **ln( )** for natural logarithm (base *e*).

Before we get there, however, in part (b), the textbook asks you to compute the ratios of the light intensity at consecutive depths. How do you do that?

- Make a new column, **ratio**.
- Give it this formula: light / **prev(light)**. The **prev( )** function in Fathom takes the value from the previous case. (The function **next( )** works the same way in the other direction.)

You will see that the first row is what we call *bogus*. That's because when there is no previous value, Fathom uses zero. We'll see how to deal with that annoying behavior later. For now, note how close all of the other values are. Since the ratios are so close together, the relationship is *très* exponential.

In part (f) you have to back-transform the equation for the line to get the exponential function and plot it on the original graph. If you used natural logs (as suggested) you will need to take *e*-to-the something. If you just use the letter **e** in Fathom's formula editor, however, Fathom will not understand (the way it does with **pi**). Instead, use the function **exp( )**. This is traditional in many computer systems. It means "*e*-to-the," so that, for example, **exp(k * depth)** means $e^{k \times depth}$.

4.6 GYPSY MOTHS. This is very much like the previous problem except that it's exponential growth instead of decay. Also, the hint the text gives is so important it bears elaborating: transform the explanatory variable **Year** to something like **YearsAfter1975**.

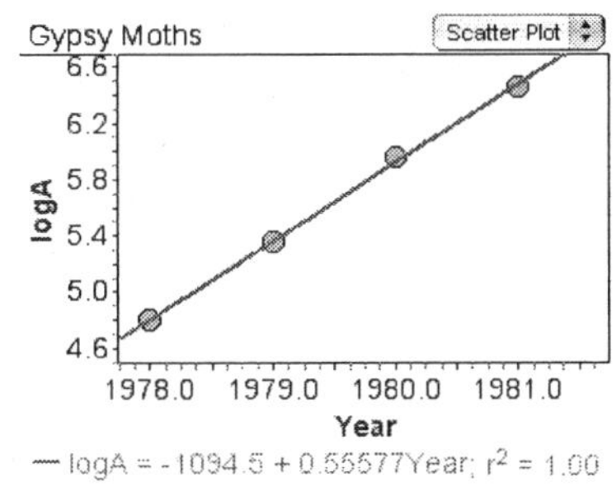

In the illustration, we used **logA = log(A)** and found a terrific linear relationship with **Year**. But what does the intercept (about –1100) mean? If you raise 10 to the power of each side, you'll find that $Acres = (10^{-1094.5})(10^{0.56})^{Year}$. The coefficient, about 10^{-1100}, is ludicrously small. Why? It's the value for **Acres** *at the year zero*. For data collected in a cluster 2000 years later, that value is likely to be absurd. If you use "years after 1975," however, the value will at least have a chance of being meaningful.

4.9 U.S. Population. In part (c) the textbook asks you to "use the data from 1920 to 2000." How do you specify a subset like this in Fathom? One way is to place a filter on the collection itself (the box of gold balls). But here is another way you will find useful:

- Plot the whole data set.
- Select the cases you want to study. They turn red in the plot.
- **Copy** them (**Edit>Copy**, CONTROL-**C** or ⌘-**C**).
- Make a new collection (drag an empty box off the shelf). 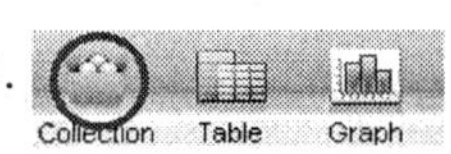

- **Paste** the cases into the collection.

- Give the collection a good name (such as **Pop 1920–2000**). Now you have a copy of the specific cases you want in a collection of their own, and you can analyze them as you would any other data set.

  ✎ Use your model to estimate the U.S. Population right now. Then go to **http://www.census.gov** and check your estimate.

4.11 BODY WEIGHT AND LIFETIME. Here is an interesting extension: add humans to the collection of mammals. Are we an outlier? How would you interpret or explain our position on the graph?

4.13 HOW MUCH WEIGHT IS TOO MUCH? Before you do the analysis, consider: Are we predicting weight from height or vice versa? This seems just like the fish in Example 4.2; can we say that weight will be proportional to the cube of height? Solve the problems in the textbook, and see if you can explain the exponent you get.

A NOTE ON POWER LAWS AND POWER TRANSFORMATIONS. You will often have to distinguish exponential situations from power-law situations. Here is one way that often works: consider whether the intercept should be zero. In straight exponential growth, there is an intercept, the "initial" value of the response variable. In a regular power-law situation with a positive exponent, (i.e., $y = Ax^p$) if x is zero, y is zero.

The Return of the Fish

At the beginning of this chapter, we used some data on fish to learn about transforming data to do nonlinear modeling. We said that the weight of fish was probably proportional to the cube of the length, $W = kL^3$. But that's also the form of a power law, $y = Ax^p$, with $p = 3$.

- Analyze the fish data (either set, the rockfish or the snook) as a power law, that is, plot the log of the weight against the log of the length, and make a linear model. Should it go through zero? What slope do you get? What does the slope mean?
- Back-transform the linear function to plot a nonlinear function on the original, untransformed data. How similar is the result to what you got before?

  ✎ Extension: Our best exponent is slightly different from 3.0. Yours may be too. What does that difference mean in terms of the shape of the fish? (Hint: What assumptions did we make in order to predict that 3 would be a good exponent?)

Extra: Nonlinear fitting without transformations

Is there any way to do all this without using (yuck) logarithms? You bet. We'll continue with the fish data. Use the **EG04_02 Rockfish** file. Suppose we think that this relationship is cubic, $W = kL^3$.

- Make a plot of **Weight** versus **Length**.
- Make a new *slider*—just drag it off the shelf—and change its name from **V1** to **K**.
- Plot a function on the graph with the formula **Weight = K * Length^3**.

❖ Change the value of **K** by dragging the pointer or by typing until the function matches the data. Looking at a residual plot helps, just as it did with linear functions.

❖ If you have a **units incompatible** error, change the units of **K** so that the units on the two sides of the equation for the curve balance.

A couple of notes:

> To find a good value for **K** in this situation, you will need fine control over small values. So re-scale the slider axis by dragging the larger numbers off the end. Another strategy is to hold down CONTROL (Windows) or OPTION (Mac) and point at the slider axis. The cursor becomes a magnifying glass. Click to expand about the point where you click.
>
> You can make any number of sliders. Try making a **p** slider and edit the function to be **Weight = K * Length^p**. Then you can adjust that slider as well. Typically, you will go back and forth among your sliders, *always looking at the residuals* as you change the values.

4.2 Relationships between Categorical Variables

The problems in this section in the textbook are really best done by hand and with a calculator, so we will not try to duplicate them here in Fathom. What we will do instead is show you how Fathom addresses the main issue: how to see relationships between categorical variables. Do the material in the textbook *first*.

Why is Fathom inappropriate for many of the problems? Because the problems begin with *aggregate* data, and Fathom really expects data to be about *individual cases*. For example, in Table 4.5, the table tells us that there are 89,000 female college students 15–17 years of age. What is the "individual" in this data set? An age group? The females? No, really there are 89,000 individuals, and each one has an age and a sex. The "89" in the table is an aggregate statistic, a *processed* number that helps describe the whole universe of college students.

It would be cool if you could write the summary table and Fathom could make the data table of individuals, but that's just not possible yet. So instead, we'll start with individuals and make the summary tables like the ones you see in the textbook. You already did some of this in the preliminary chapter.

Getting data

Let's get data about a lot of people. There's no better place for this than the U. S. Census. Note: You must have an Internet connection for this to work.[1]

❖ In a new Fathom document, choose **Import>U. S. Census Data** from the **File** menu. An inspector appears that will let you specify what data you will collect.

❖ Ignore it for now; just press **Download Data**. Fathom churns and polls and downloads 500 cases into your collection.

---

1 If you have no access, use the **0A Introductory Census Data**, in the folder for the preliminary chapter.

❖ Make a table of the data and explore it briefly; you should see that you have a lot of people from all over the country.

We are going to study the relationship between two categorical variables: **Sex** and **Marital_status**. We'll start with graphs.

Seeing categorical relationships in graphs

❖ Make two graphs: one with **Marital_status** on the horizontal axis, and the other with **Sex** on the horizontal axis. Nothing on the vertical axes. Each should be a bar chart.

❖ Click in the various bars of the two graphs; you'll see how the proportions change in the other graph. Look for differences between the sexes. The one we'll concentrate on, that appears reliably, is that *more females are widowed than males.* The illustration shows one pair of graphs that illustrate this:

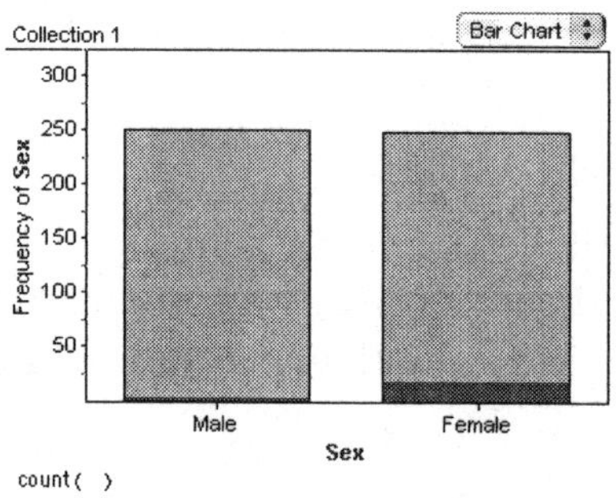

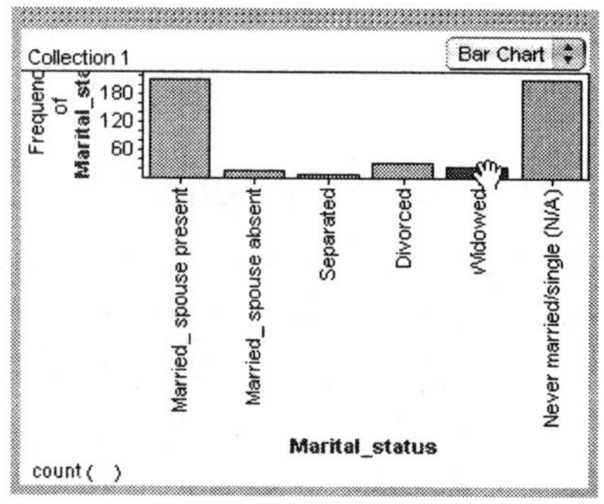

In the illustration, we've just clicked on **Widowed** in the right-hand graph. You can see that the number of selected females is a lot more than the number of selected males.

Working with multiple graphs and selecting is a great way to see relationships among categorical variables. But you can also get more information in a single graph.

❖ Drag **Sex** from the first graph to the vertical axis of the second. See what you can tell from that graph. Then drag **Marital_status** from the second graph to the vertical axis of the first. The two graphs show the same information, but the one with **Sex** on the horizontal axis seems to be comparing the sexes, while the other is comparing the various marital statuses. We call these "split bar charts."

❖ Choose **Breakdown Plot** from the menus in the corners of the graphs. The display changes to show the individuals again, bunched into squares. Here you can see the gender difference among widows:

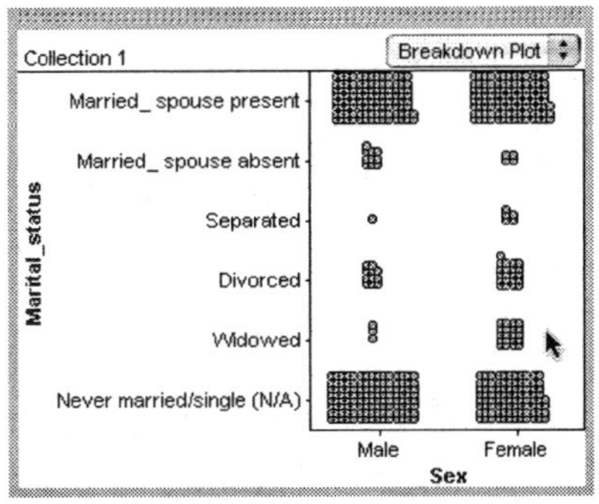

Split bar charts and breakdown plots can be useful, but for comparing groups like these, nothing quite beats a ribbon chart. Ribbon charts are a little unusual, but with practice you'll make them your friends. Here's how to make them:

- Get rid of your current graphs. Then make a new one with **Marital_status** on one axis. (We put it on the vertical axis for the illustrations.) You get a bar chart.
- Change from a **Bar Chart** to a **Ribbon Chart** using the menu in the corner of the graph. This gives you a ribbon chart, a "stacked" chart where each segment is proportional to the number of cases with that value. The axis is in percent. The names of the values may overlap a bit; don't worry about that. (See the left-hand graph in the illustration.)
- Now drag **Sex** to the *middle* of the graph. Each bar splits, and a legend appears below. Now you can see how the **Widowed** segment is mostly female and how that proportion compares to the other proportions of females in the other segments.

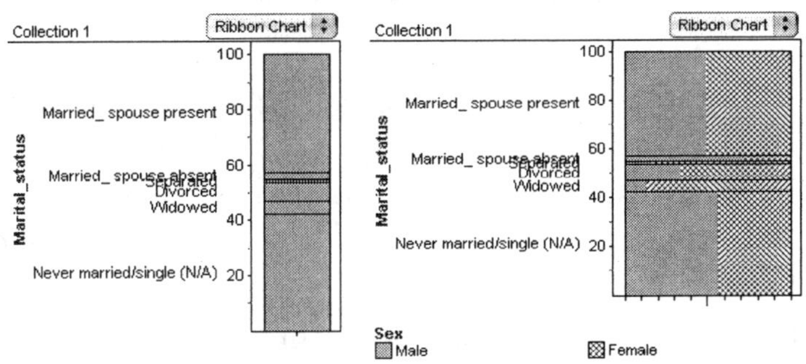

The right-hand graph in the illustration shows what you can expect. Now note: This graph shows us how the sex balance changes with marital status. To get a graph oriented the other way, do the following:

- Make a totally new graph, and put **Sex** on one axis (we used vertical).
- Drag **Marital_status** to the middle of this new graph. (You can put the "legend" on bar charts too. Some people prefer these to ribbon charts.)
- Change it to a ribbon chart. You may want to stretch the result horizontally to better see the segments.

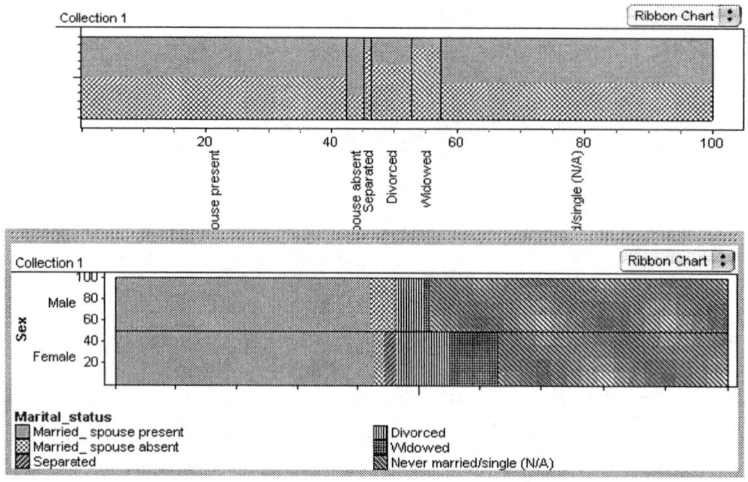

This illustration shows the two graphs, partly overlapping. We have swapped axes on the first and stretched it. The graphs show the same information, but from different perspectives. The top one shows that, among widows, a bigger proportion are females. The bottom shows that, among females, a bigger segment are widows (and, in this case, divorced).

As usual, you can point your mouse at one of these segments; if you just hover without clicking, information about the segment appears in the status bar at the bottom of the screen. If you click, you select all of the cases in the segment.

Seeing categorical relationships in tables

Graphs are great but sometimes you need the numbers. For that, the right tool is a summary table.

❖ Drag a summary table off the shelf.

❖ A summary table is like a graph. It starts out empty, and you drag variables to its "axes" to make it fill with data. But for the summary table, the "axes" are the headings at the top and left side of the table.

❖ Drag **Sex** to the top of the table and **Marital_status** to the left. You will have to expand the table to make room for everything:

Collection 1

		Sex		Row Summary
		Male	Female	
Marital_status	Married_ spouse present	106	107	213
	Married_ spouse absent	10	4	14
	Separated	1	5	6
	Divorced	11	21	32
	Widowed	3	20	23
	Never married/single (N/A)	120	92	212
	Column Summary	251	249	500

S1 = count( )

Note: We put **Marital_status** on the left because its labels were longer. Try it the other way; you'll see what we mean.

The summary table is an aggregate table like Table 4.5 in the textbook. It even shows the marginal totals just as the text describes under **Row Summary** and **Column Summary**. (Your numbers will probably be different because your sample is different from ours.)

The table shows counts. What if we want percentages? As the textbook explains (and as you know from looking at the two ribbon charts) there are two ways to compute them—by row and by column. That is, do you want to know the percent of widows who are women or the percent of women who are widows?

You need to know how to get both; here's how:

❖ Context-click on the summary table and choose **Add Formula** from the context menu.

❖ In the formula editor, enter **rowProportion**. (This is a special term in FathomSpeak.) Click **OK** and expand the table to see the results.

Collection 1

		Sex: Male	Sex: Female	Row Summary
Marital_status	Married_ spouse present	106 0.49765258	107 0.50234742	213 1
	Married_ spouse absent	10 0.71428571	4 0.28571429	14 1
	Separated	1 0.16666667	5 0.83333333	6 1
	Divorced	11 0.34375	21 0.65625	32 1
	Widowed	3 0.13043478	20 0.86956522	23 1
	Never married/single (N/A)	120 0.56603774	92 0.43396226	212 1
	Column Summary	251 0.502	249 0.498	500 1

S1 = count( )
S2 = rowproportion

Collection 1

		Sex: Male	Sex: Female	Row Summary
Marital_status	Married_ spouse present	106 50	107 50	213 100
	Married_ spouse absent	10 71	4 29	14 100
	Separated	1 17	5 83	6 100
	Divorced	11 34	21 66	32 100
	Widowed	3 13	20 87	23 100
	Never married/single (N/A)	120 57	92 43	212 100
	Column Summary	251 50	249 50	500 100

S1 = count( )
S2 = round(100rowProportion)

The formulas at the bottom (**S1** and **S2**) are keys to each of the cells. So in each cell, the count is on top and the proportion is on the bottom. Click on a formula to highlight it. You can see that 0.869565... of the 23 widows are female. To add another formula, use **Add Formula** again. But to change an existing formula, double-click it.

❖ There are way too many decimal places, and besides, we want percentages. Double-click the **S2 = rowProportion** formula.

❖ Edit it to be **round(100 * rowProportion)** as in the right-hand summary table in the illustration. Now we have percentages, rounded to the nearest whole percent.

❖ Let's see the percentages the other way: Add a new formula to the table; this time, make its formula **round(100 * columnProportion)**. You'll see something like this:

Collection 1

		Sex: Male	Sex: Female	Row Summary
Marital_status	Married_ spouse present	106 50 42	107 50 43	213 100 43
	Married_ spouse absent	10 71 4	4 29 2	14 100 3
	Separated	1 17 0	5 83 2	6 100 1
	Divorced	11 34 4	21 66 8	32 100 6
	Widowed	3 13 1	20 87 8	23 100 5
	Never married/single (N/A)	120 57 48	92 43 37	212 100 42
	Column Summary	251 50 100	249 50 100	500 100 100

S1 = count( )
S2 = round(100rowProportion)
S3 = round(100columnProportion)

Aha! Now we see that (for example) 8% of women are widows, but 87% of widows are women. Notice that Fathom also produces the *marginal percentages*.

The percentages in the middle of the table are *conditional distributions*, and they cut two different ways. **S3** shows the distributions of marital statuses "conditioned on" (that is, *assuming*) sex is "female" or sex is "male"; and **S2** shows the distribution of sex, conditioned on each marital status.

Getting education data (requires Internet access)

Enough of marital status. Let's find out if our census samples will show us whether college students show the same age and sex distribution as the dataset from the textbook. For that we need age (which we have) and education data (which we don't).

❖ In a new Fathom document (**File>New**), choose **File>Import>U.S. Census Data**. An empty collection appears, with its inspector open to the **Microdata** panel.

❖ In the **Microdata** panel, open **Attributes**, then **Person**, then select **Education**.

❖ Select **School Attendance** and **Educational Attainment**. Your inspector will look something like this illustration:

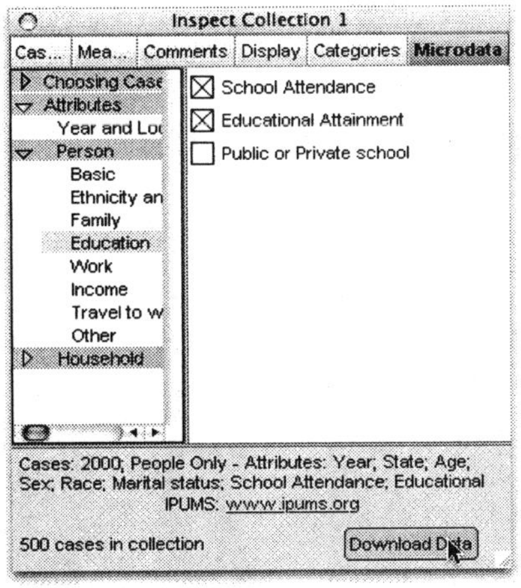

❖ Click **Download Data** to get 500 cases with the new variables. Make a data table so you can see them. (If it is empty, drag the name of the collection into the empty table.)

❖ Make a new graph of the two new variables: **Educational_attainment_recode**[2] and **School_attendance**. Figure out what you're seeing. You will see, for example, that most of the people in school have completed only elementary or middle school.

❖ Change it to a **Breakdown Plot**. There, select all of the people who are in school and have completed grade 12 or higher (so they must be in college).

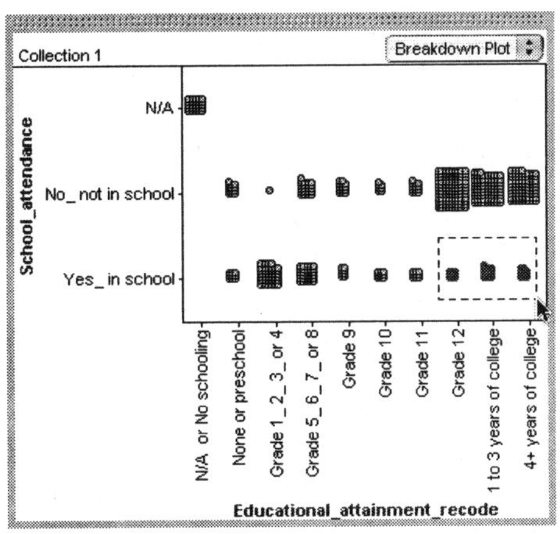

❖ Wait a second—this isn't enough people! Let's go back to the inspector for this collection. (If the inspector is closed, double-click the collection.) Go to the **Microdata** panel.

❖ This time, open up **Choosing Cases** and select **Sample Size**. There, enter 2000 and click **Download Data.**

The great thing is, all of the graphs that you have set up still work. Fathom just replaces the data, not all of your analysis.

2 Remember, you can change it to **Education** or whatever you want. Double-click the name to edit it.

The next steps will get you a collection containing *only the college students*. There are lots of ways to do this; this is only one.

- Select the (now larger number of) college attenders in the breakdown plot, as shown in the previous illustration, and **Copy** them.
- Open a new Fathom document, *make a new empty collection*, and **Paste** the cases in. Rename your collection **College students**. We had 140 of them. (You don't *have* to make a new Fathom document, but ours was getting crowded.)
- Explore the data. For example, make a histogram of **Age** and a bar chart of **Sex**, and select the women. Does the pattern of sex and age match the one in the textbook?

Recoding age

If we want the same age categories as in the textbook, we have to make them. You will often have to *recode* a quantitative variable such as **Age** into categories. Here's one way:

- Make a new variable called **AgeCat**. Right-click on its name and choose **Edit Formula**.

In the formula editor, we're going to make a **switch( )** statement. These can be confusing at first. Remember to look in **Help** later if you can't remember the details.

- In the formula editor, enter **switch(Age)**. Fathom adds a whole bunch of other stuff.

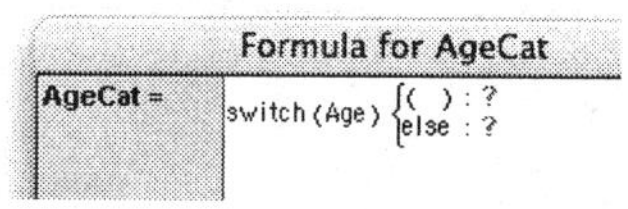

- Press TAB or otherwise move your cursor to between the parentheses at the top.
- Enter **< 18** and press TAB again. The cursor jumps to the question mark after the colon. Enter **"15 to 17"** *with the quotes.*

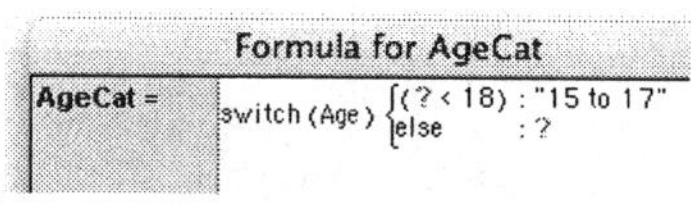

The statement so far means, "look at **Age**. If it's less than 18, make **AgeCat** equal to the text, "**15 to 17**". We don't have enough choices if we want 4 categories! We need more.

- Keep your cursor in that string. On Windows, press CONTROL-ENTER; on the Mac, press OPTION-RETURN. A new line opens up. Do that again because we need 4 branches altogether in our **switch( )** statement.
- Enter more expressions until your editor looks like the illustration. Note that pressing TAB moves you from one editable spot to another. SHIFT-TAB moves you backwards.

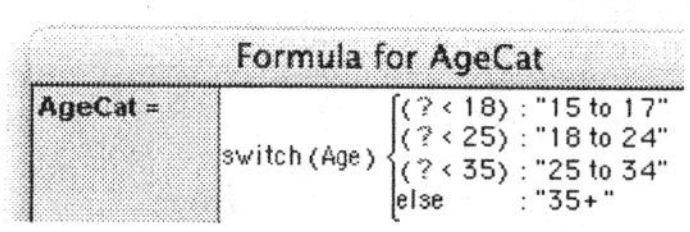

Be sure you understand what this strange statement means: Look at **Age**; then (reading from the top, to the right of the bracket) if it is less than 18, make **AgeCat** "15 to 17" (and stop); then if it is less than 25, make **AgeCat** "18 to 24". And so forth.

If **Age** was 16, **AgeCat** would *already* be "15 to 17", so it doesn't trigger that second, **<25** condition. But if **Age** is 25 or more, continue. At the end, if Fathom has not assigned **AgeCat** yet, it makes the string "35+".

This type of construction is really, really common in statistical computing. There are variations to the **switch( )** statement, and its simpler cousin, **if( )**, that we will not go into here. It is inevitable that you will mess one of these up sometime, but that doesn't matter. Just check to make sure the formula is doing what you think, and re-edit the formula until it comes out right. (How do you check? For example, plot **Age** against **AgeCat**.)

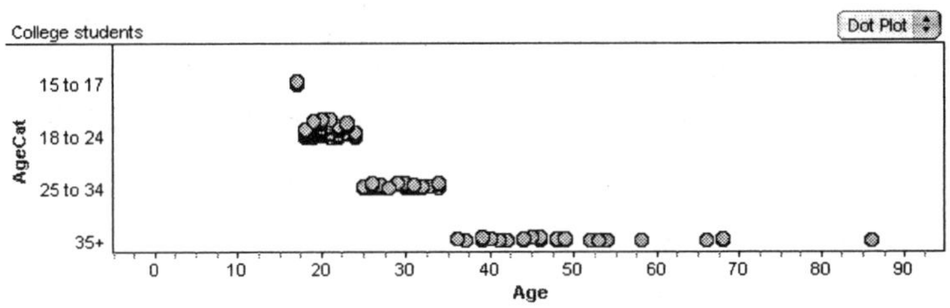

❖ Now you can analyze **AgeCat** just as you did **Marital_status**. See if you can make the table like the one in the textbook. Then get conditional distributions so you can study how sex balance changes with age.

College students

AgeCat	Female	Male	Row Summary
15 to 17	2	2	4
18 to 24	44	29	73
25 to 34	16	17	33
35+	18	12	30
Column Summary	80	60	140

S1 = count( )

College students

AgeCat	Female	Male	Row Summary
15 to 17	2 50	2 50	4 100
18 to 24	44 60	29 40	73 100
25 to 34	16 48	17 52	33 100
35+	18 60	12 40	30 100
Column Summary	80 57	60 43	140 100

S1 = count( )
S2 = round(100rowProportion)

You can see that the overall pattern from the textbook—more women than men in college, and more older women than older men—holds true for our sample of 140 college students.[3]

Example 4.15 Simpson's paradox: Do medical helicopters save lives?

Let us quote from page 300 in Example 4.15: "At first, it seems paradoxical that the helicopter does better for both groups of victims but worse when all victims are lumped together."

True, and that is the heart of Simpson's paradox. We have avoided some of the problems and examples in this section because they use aggregate data, and Fathom does best with case-by-

3 Remember that we harvested them from a sample of 2000 individuals of all ages.

case information. But this is so important, we constructed the file that will give you the right aggregate numbers for this example.

- ❖ Open **EG04_15 Simpson.ftm**. It has the data.
- ❖ Use Fathom to make tables or graphs that show, convincingly, that victims transported by are more likely to die, and yet for both serious and non-serious accidents, victims are more likely to survive if transported by helicopter.
- ❖ Devise a presentation using Fathom (or graphs and tables from Fathom) to explain to someone how this is possible.

4.3 Establishing Causation

As the textbook explains, establishing causation is tricky, especially if you can't do a controlled experiment. Fathom cannot control variables or identify confounders for you, but it can help you explore the data thoughtfully. We'll do some of that in section 5.2, coming up in the next chapter.

5 Producing Data: Surveys, Observational Studies, and Experiments

Fathom doesn't think for you, so it cannot help you design a survey well, or decide how to control for confounding variables. It can, however, help you with some of the mechanics of sampling, and it can help you assess whether confounders are a problem.

For example, the chapter in the textbook points out how important randomness is in sampling. Fathom is really good at randomness. Let's begin by seeing how it can help you construct a simple random sample (SRS).

5.1 Designing Samples

Suppose you want to pick five states at random to study.

❖ Open the file **05 States.ftm**. Our data includes the 50 state names and their populations (taken from the U. S. Census in 1990 and 2000).

❖ Right-click the collection (the box of balls) and choose **Sample Cases** from the context menu. A new collection appears, called **Sample of States**. (The balls in the box are blue instead of gold.)

❖ Make a table to show the contents of this new collection. It should have 10 states in it.

There are two things we need to do: First, of course, we have to get Fathom to pick five states instead of ten. Second, we have to make sure that Fathom won't pick the same state twice. In the illustration, for example, we have two Louisianas:

Sample of States

	State	Pop2000	Pop1990
1	Louisiana	4468976	4219973
2	Montana	902195	799065
3	Vermont	608827	562758
4	Connecticut	3405565	3287116
5	Hawaii	1211537	1108229
6	Nebraska	1711263	1578385
7	Indiana	6080485	5544159
8	Delaware	783600	666168
9	Pennsylvania	12281054	11881643
10	Louisiana	4468976	4219973

❖ Double-click the sample collection to bring up its inspector. It will open to the **Sample** panel, and should look like the illustration below left. Change it to look like the illustration on the right.

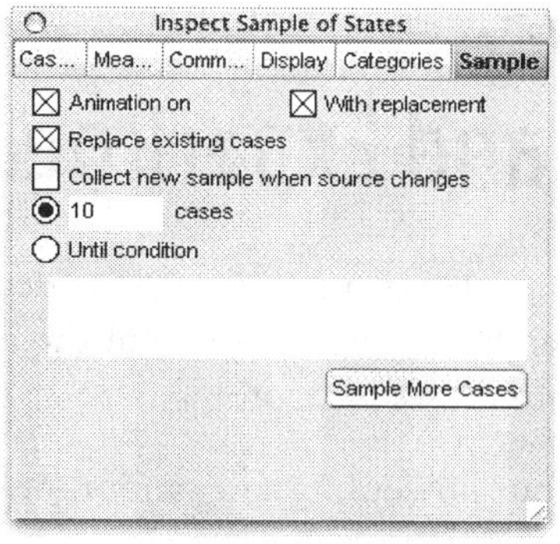

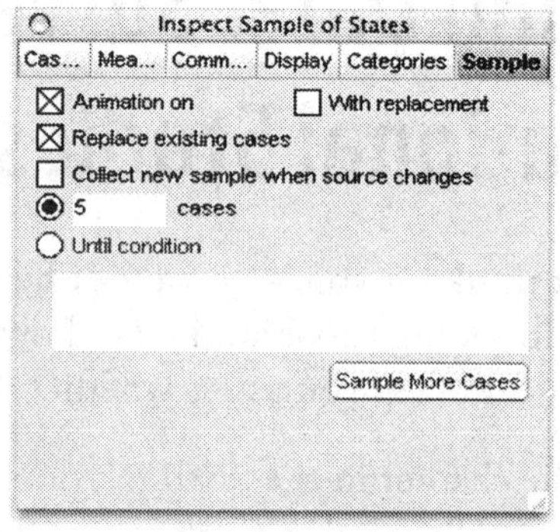

Changing the 10 to a 5 is straightforward. The other change is to un-check **With replacement**, as in the right-hand illustration. That ensures that when Fathom picks a state, it won't put it back into the pool to be picked again. With that box un-checked, we're sampling *without* replacement.

WHY WOULD YOU SAMPLE *WITH* REPLACEMENT? It turns out to be common (in probability problems, for example) to replace the objects you sample. It's so common that Fathom made it the default behavior. So remember: If you want to sample things only once, un-check that option.

- Press the **Sample More Cases** button. Fathom now picks five—without replacement—and displays them. Just press the button to get another sample.

Here are two more ways to get a new sample. You need them when the inspector is not open:

- Right-click the sample collection and choose **Sample More Cases** from the context menu.
- Expand the collection a little by dragging on an edge or corner (see the illustration below). A **Sample More Cases** button appears. Press it to get a new sample.

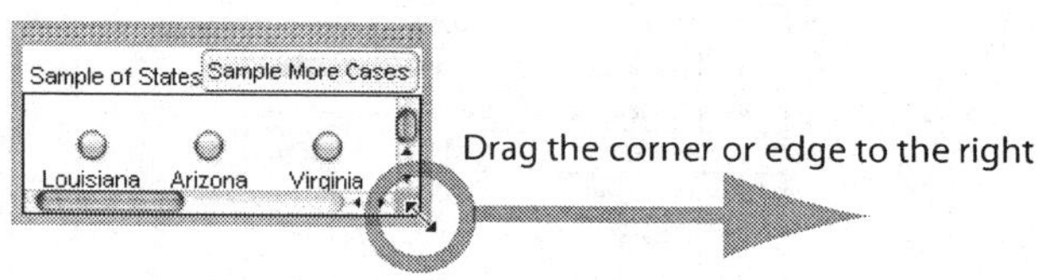

What is our population?

Our population is the states themselves. We have a simple random sample (SRS) of five states. Note that this is *not* the same as choosing five *people* at random and then asking them what state they are from. In that case, the population would be the people. If we did that, we'd expect more Californias to show up, and fewer Vermonts.

There is no easy way (yet) to accomplish that kind of sampling in Fathom. We would really need a collection with one case for each person. You can use the population data to adjust the probability that each state gets chosen, but doing so gets complicated.

Be very clear what your population is. Not doing so can taint your results. For example, suppose you want to know the average size of families. If you sample *people*, asking them their family size, you will get a bigger number than if you sample the *families*. Why? Because there are (by definition!) more people per family in large families than small. If you get a genuine SRS of the people, the big families have a bigger chance of being sampled than the small ones. (You will meet this later when you do our version of Exercise 7.13 on page 101 of this guide.)

5.2 Designing Experiments

Let's do an observational study using Fathom. It won't be a true experiment, as you will see; we'll take the data as we find it. We're not creating a treatment or randomizing, so we have no control over confounding variables. Still, we can do our best to account for them. In the course of this extended example, you'll see how you can use Fathom to deal with typical problems.

A word of advice, though: No experiment or study is truly typical. This is not an area where you can follow a clear recipe; you have to think carefully every time about the variables, what they mean, and what other effects could bring about any pattern you see in the data. So although these techniques will be useful to you, think of them as examples of the kinds of technique you will *invent*, rather than ones you memorize.

Onward! We'll begin with a conjecture—a statement we think might be true. We're going to study it to see what we can say about it:

Pregnant people are, on the average, heavier than non-pregnant people.

Makes sense, doesn't it? Note that there is really no *cause* implicit in our conjecture. It might be that people weigh more because they are pregnant; or it could be that people get pregnant because they weigh more; or there could be a whole separate mechanism through which people become both pregnant and heavy. Let's not concern ourselves with that issue just now; instead let's work to establish (or refute) the claim that pregnancy and weight are *associated*.

To do so, we are going to use some great data from the National Health And Nutrition Examination Survey (NHANES). We'll set this out as a tutorial for you to do step by step. We encourage you to explore, but always return to the "main thread" before you go on.

❖ Open **05 NHANES.ftm**. It contains records of the height, weight, race, pregnancy status, education, etc., of over 8000 people.

❖ Explore the data a little to see what you have. Start by making a table.

Our first attempt at verifying or refuting the conjecture should generally be a graph.

❖ Make a graph of **Preg**. You'll see that most people are not pregnant or that pregnancy "does not apply," indicated by **N/A**. (More about this later.)

❖ The graph shows how many people are in each category. We want to know about **Weight**. Put **Weight** on the other axis. You'll see a whole lot of dots. Change it to a box plot to reduce the clutter. (Left-hand graph in the illustration.)

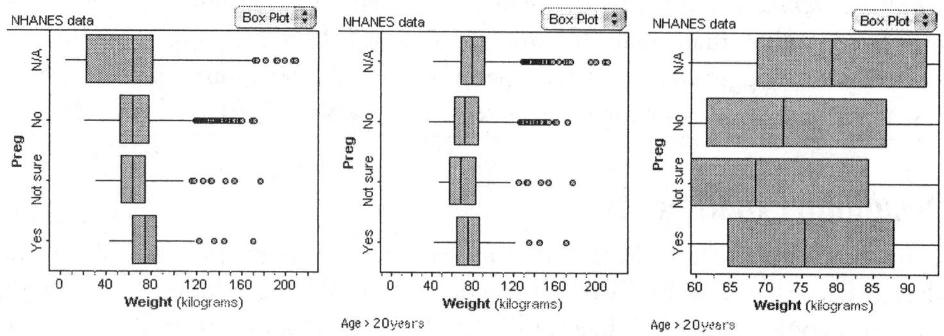

Hooray! Our conjecture is right! Pregnant people do have higher **Weight**s. Don't cheer too soon, though. As you may have noticed, we included the children in this investigation. Suppose we look only at people over 20.

❖ Right-click the graph and choose **Add Filter** from the context menu. In the formula editor, enter **Age > 20 years**. If you don't include the units, Fathom will punish you.

❖ When you click **OK**, the graph changes, reflecting the filter. Now it shows only people over 20 years of age. (Middle graph in the illustration.) If you need to edit the filter, double-click it there at the bottom of the graph.

USE FILTERS TO FIND SUBSETS. The filter is a powerful tool. Filter any Fathom object to make it show only a subset of your data. Ordinarily, a filter applies only to the object (graph, table, whatever). But if you filter the *collection*, it applies everywhere. Shortcut: CONTROL-**F** or **⌘-F**.

❖ Now it's not obvious which category has the highest median weight, so expand the scale of the graph (right graph) and see that the **N/A**'s have the highest median weight.

❖ Find at least two ways to show that the **N/A**'s have the highest *mean* weight as well.

Does this mean our conjecture is wrong? *Yes.* On the other hand, you could rightly complain that it's just because of the way the conjecture was stated. It's true that, in general, non-pregnant adults weigh more than pregnant ones—but that's because *that group includes all of the men.* What we meant to say was, pregnant women weigh more than non-pregnant *women of childbearing age.*

This is what **N/A** means; you get put in one of the "real" categories, **Yes**, **No**, or **Not sure**, only if you're a woman of childbearing age. So we should eliminate the **N/A**'s. We could use a filter again, but another strategy is to make a copy and then eliminate what we don't want:

❖ Right-click on the **NHANES data** collection and choose **Duplicate Collection** from the context menu. A new one appears, called **NHANES data_1**. Rename it something like **NHANES WCBA** (for Women of Child-Bearing Age).

- Make a graph of **Preg** from that collection. Select the **N/A**'s. Then **Cut** them (or right-click on the graph and choose **Delete Cases)**. This can take some time; be patient. Now the graph should show the numbers of **Yes**, **No**, and **Not sure**—and no more **N/A**'s.
- Let's get the mean weights a different way. See the formula at the bottom of the bar chart (left-hand illustration)? It should read **count( )**. That means that the bars show counts in each category.
- Double-click that formula to bring up the formula editor and change that formula to **mean(Weight)**. Click **OK** to see a graph like the one below right.

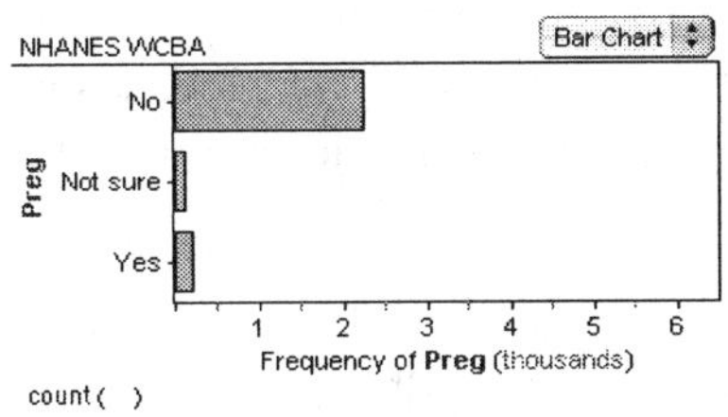

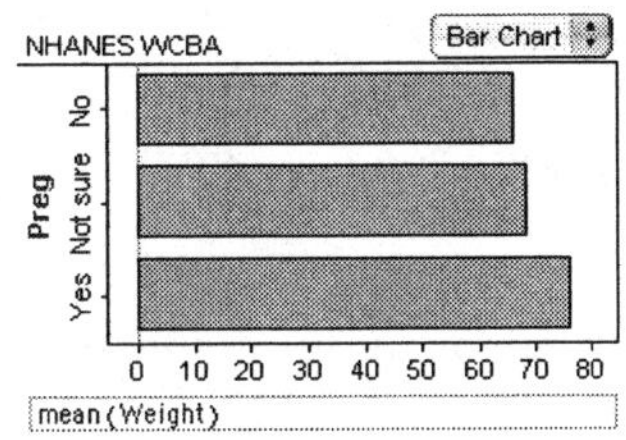

- Now we see that the **Yes** group *is* heavier, on average. Does that mean that our conjecture is true? Well, maybe. If we explore a little more we'll find out what's wrong with it.
- Make a Box Plot showing **Weight** split by **Preg**; you'll see that the median weight for pregnant WCBA's is more too.
- Duplicate the graph and replace **Weight** with **Height**. (It's in the context menu.) The two graphs appear, rescaled, in the next illustration.

DUPLICATING GRAPHS. This can save a lot of time; if the graph was hard to make, just duplicate it and slide in a different variable. The command is in the context menu, or select the graph and use the shortcut CONTROL-**D** or ⌘-**D**.

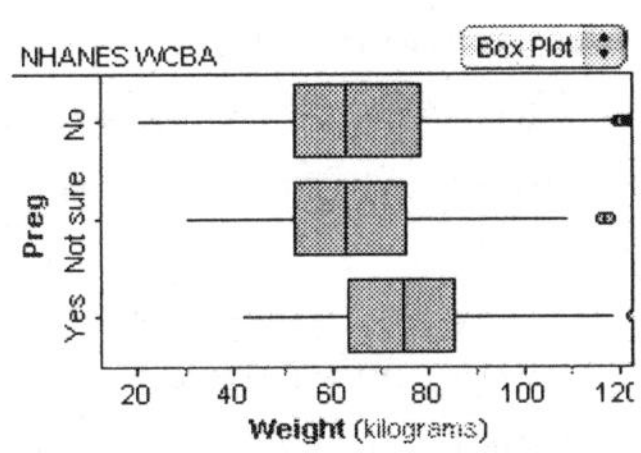

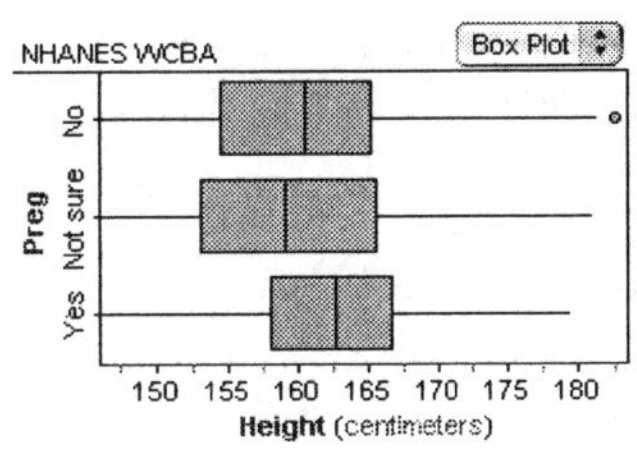

Oh, no! Pregnant women are *taller*, too! What could that mean? There are several possibilities. Is there some reason taller women are more likely to be pregnant? Surely pregnancy doesn't *cause* height gain. A little more exploration shows us what's up. Plot **Preg** against **Age**, and you'll see that many of the **No** people are very young—as young as 8![1] Once again, the conjecture seems true, but unfairly so. One more time, the mix of ages has confused the relationship between pregnancy and weight.

1 You'll see that many are older, too; in fact, if you go back to **mean( )** right now instead of **median( )**, you'll see that the mean age of **No** is about the same as for **Yes**.

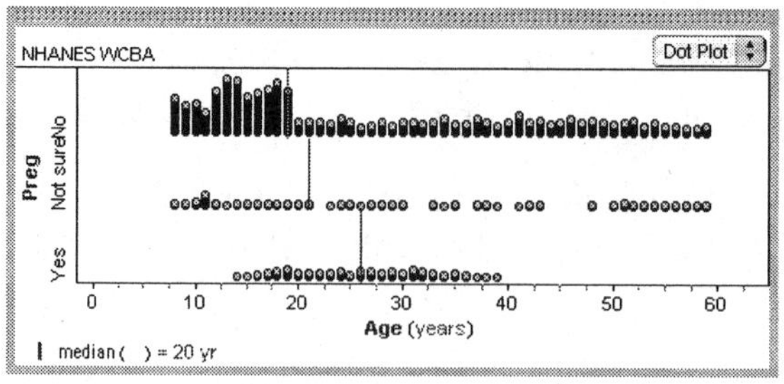

Put another way, because fully grown women are more likely to be pregnant, we can say that age is associated with pregnancy. Age is also associated with weight (think: why?). As a result, pregnancy is associated with weight. We say that age is a *common cause*, associated with both weight and pregnancy; even if pregnancy didn't make you heavier, we would see association just because of age.

But what we *really* meant in our conjecture was that the same kinds of women will be heavier when pregnant than when they're not. So let's filter our graph to show only women between the ages of 20 (we'll assume 20-year-olds are fully grown) and 40 (the top age of pregnant women in our graph).

❖ Select the **Preg** box plots of **Weight** and **Height** and **Add Filter**s with the expression **InRange(Age, 20, 40)**. After you rescale, you'll see graphs like these:

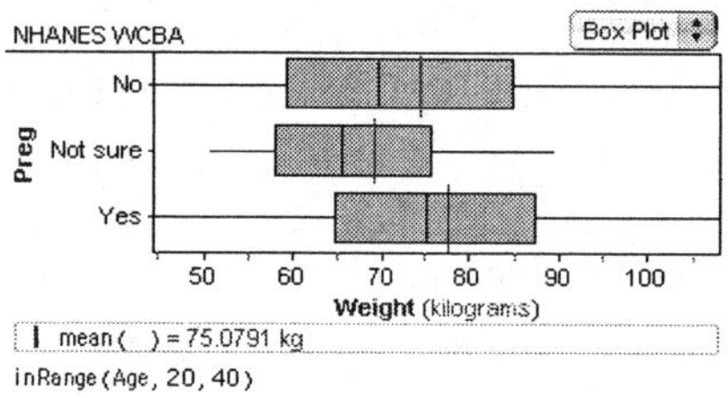

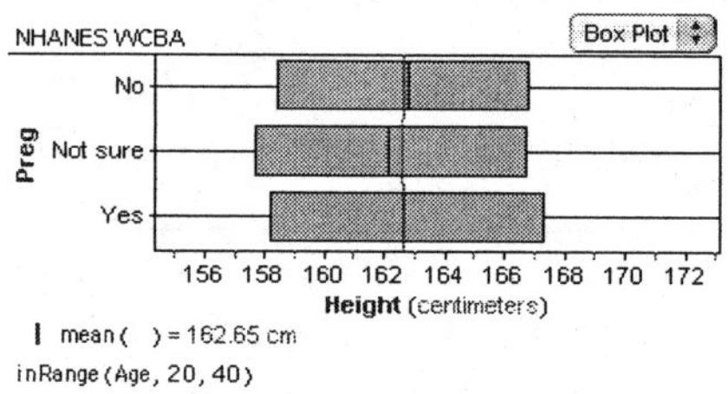

We have used **Plot Value** to show the means as well as the medians you get from the box plots. In the graphs, we can see that the pregnant, **Yes** women are on average a little heavier than the others, *but no taller*. Whether it's statistically significant is a topic for another chapter, but in any case, the difference is not dramatic.

The astute reader might go even further and say that the key comparison is to take women in the 20–40 age range and compare **weight** to **height**. That is, see if pregnant women are heavier than non-pregnant ones *of the same height*. How can we do that?

❖ Make a scatterplot of **Weight** versus **Height** and drop **Preg** into the middle of it.

❖ Right-click on the graph and pick **Least-Squares Lines**. Three lines appear, one for each value of **Preg**.

The equations at the bottom are identified by the symbols in the legend; for example, the circle means **No** (not pregnant). If you click on one of these equations, its line highlights in the graph by getting thicker. In the illustration, we have selected the **No** line. It's the lower of the two upward-sloping lines; the upper one is **Yes**, the pregnant women. It is troubling that

Not sure has a negative slope; we would not want to draw definitive conclusions without understanding that effect—or at least having a good idea why it occurs.

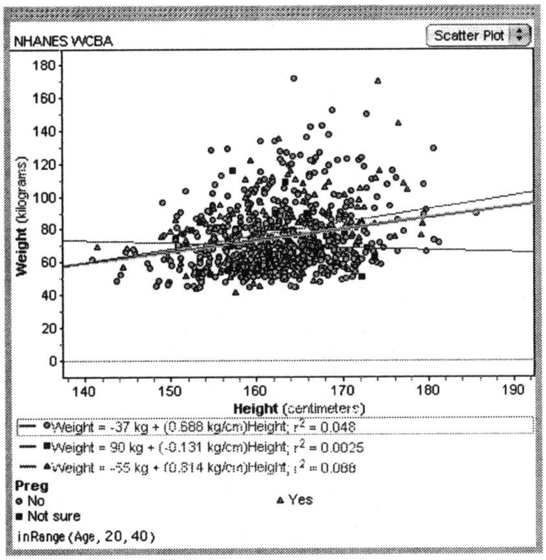

Even though our investigation of weight and pregnancy is not an experiment—and to perform one would be unconscionable—much of what we did resembles the data analysis you do with a complicated experimental study. As in our observational study, despite trying to control all confounding variables, you will never fully succeed, and you will have to do your best with the information you have.

Lessons

- ❑ When you create a conjecture, figure out what population it applies to. You may have to change your mind when you're in the midst of data analysis, but try anyway!
- ❑ Lurking variables are everywhere. In our little tutorial, at first, pregnant people were the heaviest, but that was because we had included all **Age**s. Then they were the least heavy because we included both **Sex**es. Then **Age** bit us again. Finally, we tried controlling for **Height**.
- ❑ Any data source comes with assumptions. The NHANES data considers 8-year-old girls as potentially pregnant, and records data accordingly. To give another example, the Census says "Hispanic" or "Latino" is not a race, it's something else in addition to race. These are neither good nor bad, but they make you think about what you're really trying to show.
- ❑ Recording data requires that you make all sorts of decisions. In published data, you may need to know how the researchers decided what to put down. The code **Not sure**, for pregnancy, for example, can occur in a number of ways. In our investigation it didn't matter, but if you really wanted to know, you should check the documentation that comes with the study. Look in **http://www.cdc.gov/nchs/Default.htm**—or google NHANES. From there, you can make your way to the detailed

documentation of how every data item was determined. In the case of pregnancy, it's a combination of responses given in an interview and the results of a urinalysis—but there are various restrictions depending on the age of the subject, whether she has started menstruating, etc.

- ❑ Any data source is affected by its purpose, and that might affect conclusions you draw from it. For example, the original researchers may have biases, or some agenda they're pushing. But a dataset's purpose also affects its design, and that can create unforeseen problems or opportunities.

Statistical Weight

The NHANES researchers were interested in a broad picture of people's health. Because children are growing and because childhood health is so important, the sample contains a disproportionate number of young people, that is, NHANES uses a *stratified sample.* The age distribution you get, therefore, isn't the real age distribution. At left, NHANES data. At right, U.S. Census:

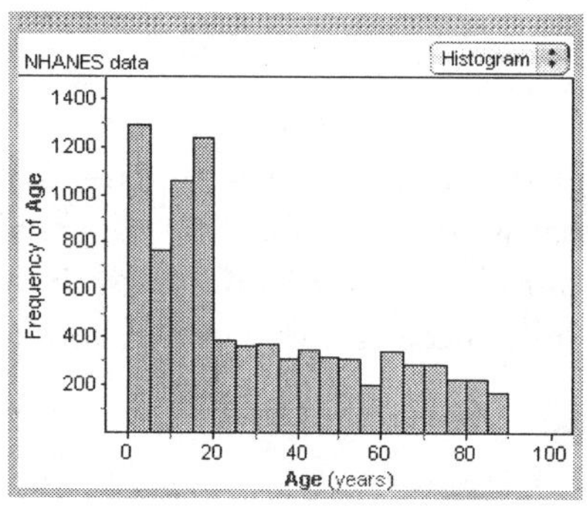

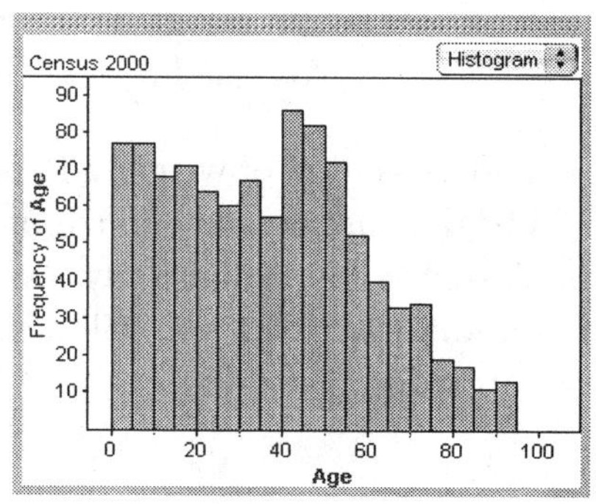

This sampling design means that the left-hand graph is *not* the age distribution the U.S. public. To account for this, the NHANES data includes a *statistical* "weight" in the variable **weight_interview**: a number that indicates how many Americans this subject stands for. Computing these weights is messy, as you might imagine. But so you get an idea, here is a scatterplot showing 100 weights as a function of age and race:

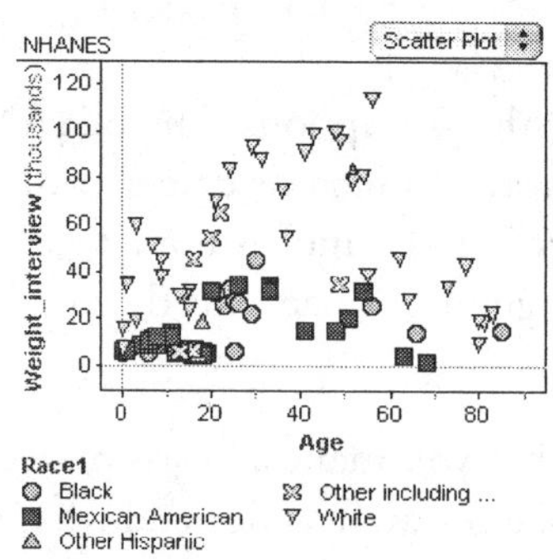

You can see that each middle-aged white person stands for as many as 100,000 people, while other races are generally under 40,000, and many children—especially minority children—stand for only a few thousand. There is no easy way in Fathom 2.0 to account for these weights; check **http://www.keypress.com/fathom** to see when Fathom learns how to handle them.

Fathom and Randomization

Although Fathom can't design an experiment for you, it can help you with one very important task: randomization. We saw earlier in this chapter how to sample (without replacement) from a population. You can use that technique to make sure that what you want randomized truly is, by making a list and scrambling it.

To do that, sample, without replacement, a number greater than or equal to the size of your collection. If you do that, your result is a collection *in random order*.

- ❑ For example, do you need to split your 30 subjects into two equal groups? Randomize your list. Then put the first 15 in one group and the last 15 in the other. If the list is in random order, you can divide them using that simple rule.
- ❑ Suppose you're worried that answers on some survey questions will influence others. Randomize the order of the questions.
- ❑ Or you're doing a Coke-Pepsi taste test with 100 people. Order matters. So you make a collection with 50 **Coke** and 50 **Pepsi**, scramble it, and use the list (secretly) to determine which drink each subject gets first. (You could also secretly flip a coin, but the "scrambling" technique ensures that in the experiment you have 50 Cokes and 50 Pepsis going first.)

This business of scrambling is so important that Fathom has a separate command for it. In the collection's context menu, you'll see **Scramble Attribute Values**. You can experiment with it to see how it works. We'll return to that command in Chapter 13.

6 Probability and Simulation: The Study of Randomness

In this chapter, you study the basics of probability and randomness. In a typical classroom activity, you might roll dice and record the results. Unfortunately, this can take a long time. The class experience can become more about rolling the dice than about the mathematics.

Using the computer has two obvious advantages:

1. It will quickly and tirelessly repeat boring procedures so you can spend your time with the results instead of rolling the dice; and
2. It will help you keep records of what happened.

There is another, more subtle advantage for learning. With the real materials, you're always switching between tools, putting away the dice and getting out the coins. When you use the computer to simulate the dice and the coins, and to do the analyses, you can better see what is the same about the phenomena, and be less distracted by differences in the equipment.

We will use Fathom to do these simulations. In this chapter, we'll use its random-number capabilities to simulate simple phenomena; at the same time, we'll type results into Fathom to record and study what happened. In later chapters, we'll learn how to get Fathom to simulate more complicated situations and to do more and more automated record-keeping.

Activity 6A Austin and Sara's game

In this activity from the textbook, Austin picks two random whole numbers from 1 to 5, and adds them. Sara picks a random whole number from 1 to 10. Austin wins if his total is larger than Sara's number. What proportion of the time does Austin win?

Let's see how to simulate playing this game many times using Fathom. To choose a random whole number, we will use a Fathom function called **randomInteger( )**.

❖ Make a new table with four attributes, **Austin1**, **Austin2**, **AustinTotal**, and **Sara**. Give them these formulas: (Right-click on the column heading—the name of the attribute—and choose **Edit Formula** from the context menu.)

Austin1	randomInteger(1, 5)
Austin2	randomInteger(1, 5)
AustinTotal	Austin1 + Austin2
Sara	randomInteger(1, 10)

❖ Give the table 200 cases. (Right-click on the table and choose **New Cases** from the context menu that appears.) Now you can see the scores for 200 games.

❖ Let's have Fathom tell us whether Austin won. Make a new column called **AustinWins** and give it the formula **Austin > Sara**. The results of the calculation are **true** or **false**. Graph **AustinWins**; you will see a bar graph.

BOOLEAN EXPRESSIONS. The formula **Austin > Sara** is *Boolean*; it's about logic, not numbers. You can combine Boolean expressions using Boolean *operators* such as **and**, **or**, and **not**. Boolean expressions are either **true** or **false**.

❖ Make a summary table as well, and put **AustinWins** on the summary table. That will tell you how many times he won out of 200 (the **true** values).

Those 200 cases give us some idea whether our prediction is true. But let's play more games and see what happens. We have two choices: increase the number of cases in our collection, or redo those 200 cases over and over. The former is easy. So we'll do the latter, because we'll learn some new skills—and get some more information.

We want to record the results of these 200 games. To that end, we'll make a new data table.

❖ With nothing selected, make a new case table with two attributes: **nWins** and **nGames**. Put the number of wins in **nWins** and the number of games (200) into **nGames**.

MY NEW TABLE ISN'T EMPTY! If you have something selected, a new table will fill with the existing data. Click in a blank area to deselect everything and try again.

❖ Rerandomize. The 200 games will change. To rerandomize, press CONTROL-**Y** (Windows) or **⌘-Y** (Mac). (You can also find **Rerandomize** in the **Collection** menu.)

❖ Record the new **nWins** in the new table. Repeat until you have 25 cases in the new table—the records from a total of 5000 games.

❖ Study the distribution of Austin's wins. If you wish, make a new column for the *proportion* of times Austin wins in each set of 200 games. Make graphs, get statistics, and make an estimate: What proportion of the time would Austin win if the game were played many more times? (That is, what do you think is the *theoretical probability* that Austin wins, based on these 5000 games?)

❖ Now consider the game from Sara's perspective. Make an estimate: Based on what you have seen so far, what do you think is the proportion of times Sara will win if the game were played many times? Let's find out.

❖ Make a new column, **SaraWins**, and give it an appropriate (Boolean) formula. Do the same analysis as you did for Austin and calculate the overall proportion of times she wins. Are you surprised?

❖ If you want to modify your estimate, what's your new estimate?

Later in the chapter, you'll calculate the *theoretical* probability and compare it to these *empirical* results.

6.1 Simulation

Simulations will help you with the big statistical ideas in the last half of this guide. Fathom is really good at simulation, and there is a lot to learn. These are the fundamental ideas so far:

- Make random integers with **randomInteger( )**. The numbers inside the parentheses are the smallest and largest numbers you get, as in **randomInteger(1, 10)**. You will learn more random number formulas throughout the next few chapters.
- Make repeated trials by making new cases (new rows) in your table. (**New Cases...** from the context menu.)
- Use Boolean expressions appropriately in formulas for results.
- Look at results using graphs and summary tables.
- Record repeated sets of results in a new case table (which creates a new collection).

Activity 6B Is this discrimination?

Follow along in the textbook; here you will construct the simulation in Fathom.

We are going to simulate 10 people being laid off, and we will assume that they are randomly chosen from among the population of salespeople. We're told that only 24% of all sales people are 55 or older. We want to know whether it is unusual for 6 of the 10 to be in the older group.

We begin by setting up the sample.

- Make a new table. Give it one variable, **older**, and 10 cases. Rename the table **Layoffs** instead of **Collection 1**.
- Give **older** the formula **randomInteger(0, 99) < 24**. Now you see the age categories of our 10 layoffs. That way, on average, 24% of our cases will have **older = true**.
- Make a summary table and drag **older** to it (top or side, it doesn't matter). Now you can see how many of the ten were older. This saves you the work of counting them.

With only one set of 10, we cannot tell whether 6 is unusual. So we will do this 20 times and record what happens in a new table.

- Make sure nothing is selected, then drag a new case table off the shelf. Give it one attribute, **nOlder** (for number of older workers laid off). Rename the table **Results** instead of **Collection 2**.

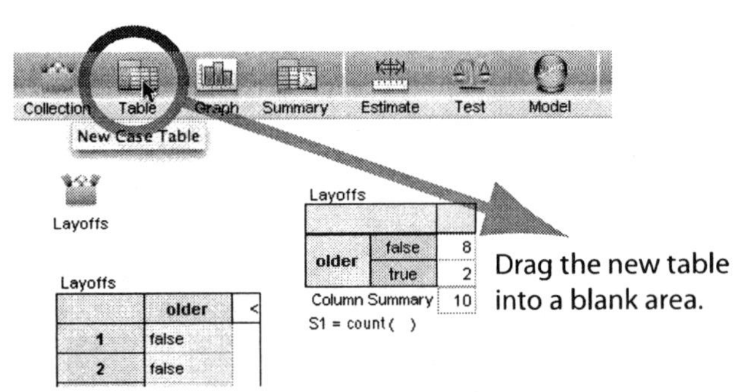

- Rerandomize to get a new set of layoffs. Record the results in the results table. Continue until you have 20 results.

Note. To rerandomize in this case, you can't just press CONTROL- or ⌘-**Y**. Right-click on the **Layoffs** table and choose **Rerandomize** from the context menu. (Fathom doesn't know which collection to rerandomize unless you tell it.)

❖ Graph **nOlder**. Use this to do step 4 in the textbook. Where should the center of this distribution be? Where is the center of your distribution?

❖ (For step 5) Calculate the relative frequency that 6 or more of the 10 laid-off salespeople are 55 or older.

❖ (Step 6) Accumulate your results with your classmates to get a better result.

✎ What's "better" about a result using more classmates?

Example 6.8 Carly Patterson, Olympics gymnastics gold medalist

In this section, we'll simulate Carly's performance using Fathom.

❖ Make a new table and give it five columns: **vault**, **parallel**, **beam**, **floor**, and **total**. Give the table 100 cases and rename it **Carly Simulation**.

❖ Give the five attributes these formulas:

vault	**randomNormal(9.314, 0.126)**
parallel	**randomNormal(9.553, 0.122)**
beam	**randomNormal(9.461, 0.203)**
floor	**randomNormal(9.543, 0.099)**
total	**vault + parallel + beam + floor**

This simulates 100 all-around scores based on the assumption that her scores will follow the same pattern as her previous work, as described in the textbook. The function **randomNormal( )** gives us a Normally distributed random variable. The first number inside the parentheses is the mean; the second is the standard deviation.

❖ Plot **total**. Use **Plot Value** to show the mean plus one standard deviation (**mean( ) + StdDev( )**) and the mean plus two standard deviations. Plot her final score, 38.387. (Just use **Plot Value** again.) Also, plot Svetlana Khorkina's score of 38.211—the score Carly had to beat.

❖ Rerandomize as many times as you like. See how Carly's actual score compares with her simulated scores.

✎ How unusual was her score? About what percent of the time would she make that score or higher given our probability model?

✎ About what percent of the time would she beat Khorkina and get the gold, given our model?

Exercises

Many simulation exercises in this chapter are perfectly suited to random-number tables, but require more Fathom features than you may know about yet. If you want to forge ahead, by all means do, and share your discoveries with your classmates. We'll make some quick comments about each one; you decide how to do them.

6.7 A GIRL OR FOUR. This is easy to simulate if you go ahead and simulate four children for every family and just ignore any children after the first girl. After all, the result is only whether there is a girl or not.

- ❖ Create a table with five attributes, **kid1**, **kid2**, etc., and **result**. Set it up so each kid is random with **0** for a boy and **1** for a girl.
- ❖ Figure out a Boolean expression for **result** so that it's true if there are any girls.

6.8 WORLD SERIES. This is tricky to simulate right now if you want to deal with stopping the series when one team wins 4. However, if you always simulate seven games, you can check to see on which game one league wins its fourth game. That will give the same result. Think about the formula **randomPick("AL","AL","AL","NL","NL")** as a way of choosing the winner. Record your results in a separate table.

In Chapter 8, you'll learn how to have Fathom collect those results for you so you can do this hundreds of times. You will use **sample until** to stop at 4 wins, and **Collect Measures** to create the distribution of series lengths.

6.9 TENNIS RACQUETS. This is another problem easier to solve using random digit tables than with the computer for now. As in the World Series problem, the trouble is getting the computer to simulate an unknown number of events. We learn a great way to do this in Example 8.15, starting on page 119 of this guide.

You can get a pretty good model, however, by simulating 25 or more matches and seeing how many he has to go until the first broken racket in each series. Record those numbers in a new collection for analysis.

6.10 SIMULATING FAMILIES. No problem! We know that each family will have 4 children, and that each child has a 50-50 chance of being a boy. If you use 0 for boys and 1 for girls, you can use a summary table to calculate the "sum of the children" in order to count the girls. Again, record the results for the families in a new collection.

6.11 THE AIRPORT VAN, I. This one you can do as well. How many cases should be in your first collection? (It's either 9 or 7. Which? Why?) How will you simulate whether a person shows up? Then, for each run of the simulation, record how many people show up altogether. Put those numbers in a new collection where you can graph them and analyze them.

6.12 SHOULD I GUESS? You need a case for each question. You also need an attribute for Jack's performance on each question (perhaps **JackIsRight**, which will be a Boolean—**true** or **false**), determined randomly, and another attribute for his **score** on that question, which is based on his performance.

If you use a Boolean expression for Jack's performance, a good expression for **score** might begin "**if(JackIsRight)**"... If you type that much in, Fathom will make an **if( )** structure you should be able to use easily.

You can do the whole thing using one attribute if you wish. It just takes a more complicated formula.

6.13 GAME OF CHANCE, I. How do you make the simulation? The easy way is to get an actual deck of cards. What makes it tricky on the computer or calculator is that if the first card is a heart, the probability of the second card being a heart is not ¼.

We will use Fathom's sampling mechanism to deal some cards from a virtual deck, and see how to deal with this problem.

❖ Fortunately, your CD comes with a deck of cards already made. Open the file **EX06_13 DeckOfCards.ftm**.

❖ Right-click on the **Deck** collection (the thing with all the cards in it) and choose **Sample Cases** from the context menu. A new collection appears, called **Sample of Deck**.

❖ Drag a corner of that collection to open it. You'll see it has 10 cards in it.

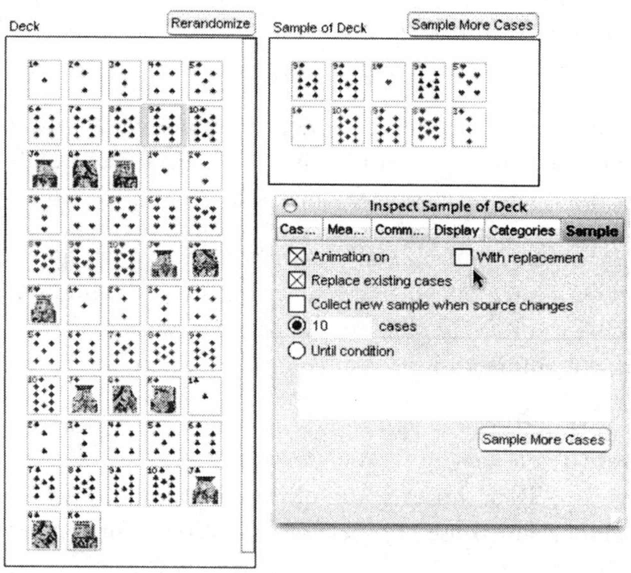

Until you un-click **with replacement**, you can get duplicate cards. In the illustration, for example, we have two nines of spades.

- This is very important: *double-click that sample collection to open its inspector.* There, *un-check* **with replacement**. If you don't do that, it will be as if you put each card back after you dealt it; it would be possible to deal the same card twice in the same hand, making the simulation inaccurate. Your window should look something like this:
- Now you can close the inspector.
- Click the **Sample More Cases** button to get another deal. Look at the first two cases to see if either is a heart.
- Make a new table to record results. Finish the problem.

Note. Change the **cases** number in the inspector from **10** to **2** if you want to deal only two cards. Change it to **5** if you want to play poker…

6.14 GAME OF CHANCE, II. You know enough to do this entirely on your own. Check back over some of the earlier examples if you have forgotten how.

6.15 THE BIRTHDAY PROBLEM. This is a famous and surprising probability situation. There are several ways to simulate this using Fathom. Here is one:

- Make a case table with 23 cases and one attribute, **day**. Give **day** a random (integer) value from 1 to 365.
- How do we know if there are duplicates? Make a summary table. Drop **day** on it. You'll see the mean of **day**.
- We are not interested in the mean, so double-click **mean( )** at the bottom of the summary table to edit it. Change it to **uniqueValues( )**. When you close the editor, the table will show the number of unique birthdays.
- Rerandomize to get a new class of 23 people.
- Record your results in a new table, and finish the problem.

  ✎ How does knowing the number of unique values help?

6.16 BATTER UP! As in Exercise 6.14, you already know everything you need.

6.17 NUCLEAR SAFETY. Again, you know enough already to make this simulation. Note that you will need separate attributes for System A and System B, and a third attribute for the combination. All of these could be Boolean.

6.18 SPREADING A RUMOR. Actually simulating the whole phenomenon is quite hard, but as the textbook suggests, you can use Fathom to help by generating the random numbers.

6.19 SHAQ. Use Fathom to simulate the sequence of 100 shots, then look at the results to get the longest run. Type that result into a new table. If you're adventurous, look up the Fathom function **runLength( )** and see how it works. You can use it to figure out the longest run for you. (Use the expression **max(runLength( ))**; where would you put it so you could see the result?)

6.20 SIMULATING AN OPINION POLL. You have all the tools, but here are some hints:

- Make a table to represent the poll.
- Change the number of women in your poll by changing the number of cases. Add cases by choosing **New Cases...** from the collection's or table's context menu. Remove cases by selecting them and choosing **Delete Cases** from that same context menu.
- Use a summary table to figure out the women's responses.
- Use a separate collection to record your results. Don't forget to graph your results to see any patterns.

6.2 Probability Models

Activity 6C The spinning wheel (Step 5)

This section tells you how to do the Fathom simulation, which takes the place of the calculator version in the textbook.

- Start with a new table and name it **wheel**. Give it 200 cases. Make four attributes, named **digit1**, **digit2**, **digit3**, and **match**.
- **Digit1**, **digit2**, and **digit3** all have the same formula: **randomInteger(1, 3)**. The **digit** columns fill with numbers.
- We need a formula for **match** that will tell us if there is a match or not. This will be a Boolean formula, that is, it will have a **true** or **false** result. Enter **match**'s formula to read **match** = **(digit1 = 1) or (digit2 = 2) or (digit3 = 3)** (as shown).

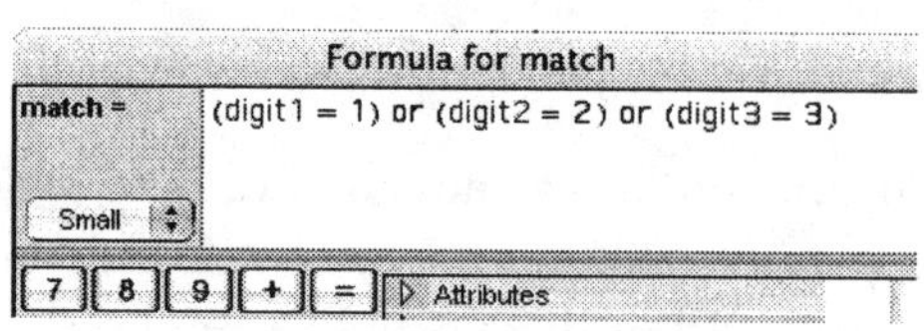

NOTE. Use the **or** button on the calculator to make the **or** operator.

✎ Why did we use **or** instead of **and**?

- Make a graph of **match** and also put it on a summary table so you can see the counts.
- Rerandomize to see how the model varies.

Activity 6D Proportion of heads versus count of heads

As an alternative to the applet, use the Fathom file **6D Proportion vs Count.ftm**. To start over, select all of the cases in the **Measures** collection and delete them. (Don't press DELETE. Choose **Delete Cases** from the **Edit** menu.)

Exercises

Even though the exercises in this section are not specifically designed for the computer, here are some comments:

6.26 HOW MANY TOSSES TO GET A HEAD. This is an interesting problem; you will learn how to simulate things like this in Chapter 8 when you study the geometric distribution.

6.28 PROBABILITY AND POKER HANDS. Use what you learned in Exercise 6.13 using the file **EX06_13 DeckOfCards.ftm** to build a rudimentary poker simulator. When you make a **Sample of Deck** collection and open its inspector, change the number of items to sample from 10 to 5. (Alternatively, use the two "rows" as two hands.) Be sure the sampling is *without* replacement.

Now you can deal many hands and record, for example, how often you get a pair.

The first step in automating the process of evaluating poker hands is to use the **uniqueValues( )** function as you did in Exercise 6.15, THE BIRTHDAY PROBLEM.

Example 6.15 Benford's law: Applying probability rules

The textbook suggests that Benford's law applies to county populations. Let's see! Open **EG06_17 Benford's Law**, which contains the populations of the 3141 U.S. counties from the 2000 Census. We have used a Fathom function to find the first digit of each population.

- ❖ Explore the data set to see what it has. For example, figure out which state has the most counties. Calculate the population of the whole United States. Find out what function we used to get that first digit.
- ❖ Now, on to Benford. Make a new summary table. Drag **firstDigit** to the *side* of the table, but *before* you drop the name on the table, *hold down the* SHIFT *key.*
- ❖ What happens if you don't hold down the SHIFT key? Try it.
- ❖ Use the numbers to see if the county populations really do obey Benford's law.

Challenge: Benford division

Benford's law suggests that numbers from real-life situations often follow a pattern so that the first digits tend to be smaller than if they were chosen randomly. In particular, lots of numbers—over 30% of them—tend to begin with **1**.

This is not true of numbers chosen from a uniform distribution such as that produced by the Fathom functions **randomInteger( )** or **random( )**. For those, the first digits are uniform as long as you choose a suitable range. (If you use **randomInteger(40, 49)** to generate numbers, you will get a lot of **4**s for first digits.)

However, if you take two random numbers that give uniform first digits and *divide* them, the resulting quotients start with **1**s more than with any other digit. In fact, the distribution of digits resembles Benford's Law.

The question is, does the distribution actually match Benford's Law pretty well, or does it just resemble it superficially? Devise a simulation, build it, and find out.

Once again, the Benford proportions:

1	2	3	4	5	6	7	8	9
0.301	0.176	0.124	0.097	0.079	0.067	0.058	0.051	0.046

6.3 General Probability Rules

This section is largely theoretical, so there are fewer opportunities for us to use Fathom. But we can make extensions to examples and exercises that will let you apply your understanding of the rules of probability.

Independence and the Census

You can explore issues of probability and independence using Census microdata. Here is an example activity to follow.

- Open a new document and choose **File>Import>U.S. Census Data...**
- In the inspector that appears, under **Choosing Cases>States**, pick **Alabama** and **Arizona**.
- Download 500 cases (this is the default when you press **Download Data**).
- Answer the following questions. If necessary, re-download data to get a different sample.
  - Estimate the probability that a person in your sample is from Arizona.
  - What is the probability that a person in the sample is black?
  - What is the probability that a person is black *and* from Arizona?
  - Estimate the probability that if a person is from Arizona, that person is black. Write this using the notation of conditional probability. Explain why this is different from the previous question.
  - Being black is *not* independent of which state you live in. Answers to the last two questions suggest this. Using data in this file, and the probability rules you have learned, give quantitative evidence why not.

On the other hand, there is not as strong an association between **sex** and state. That is, being female is "more independent" of what state you're in.

✐ Do analogous calculations about **sex** to show whether sex is independent of what state you live in.

You probably cannot say for sure that **sex** is independent of state. Real data are messy.

✐ When we were calculating theoretical probabilities, we used independence all the time without any trouble. Why does using real data make a difference?

Now design an activity of your own that gets at the same topics.

❖ Find a different Census download that will let you show that two attributes are not independent. You could compare states, or sexes, or Census years, or something completely different. You can look at race, or age, or income, or job, or whether people carpool—the list could go on and on. Show a lack of independence and write it up.

❖ Whenever you explore some association between two variables, speculate as to the *reason* these two attributes are not independent. They are *associated*—can you think of why? Be sure to identify your speculation as such, and keep it separate from the data!

7 Random Variables

In this chapter, you will learn more about how to set up probability simulations in Fathom. We'll step you through some of them, but you will be increasingly on your own.

To make simulations, you will use random-number functions. Fathom has several of these, all listed in the function browser under **Functions>Random Numbers**. There are more than you will ever need. Here are examples of some of the most useful:

random( )	returns a number between 0 and 1.
random(10)	returns a number between 0 and 10.
random(–2,7)	returns a number between –2 and 7.
randomInteger(1, 100)	returns an integer between 1 and 100 inclusive.
randomPick(1, 2, 4)	returns 1, 2, or 4, at random.
randomPick("F", "M")	returns the string F or M, randomly.
randomNormal( )	returns a number from the *standard Normal* distribution, that is, mean = 0.0 and standard deviation = 1.0.
randomNormal(mu, sigma)	returns a number from the Normal distribution with a mean of **mu** and a standard deviation of **sigma**. In this case, **mu** and **sigma** are probably sliders.

Activity 7A The Game of Craps

One element of this activity is simulating two dice. We'll use this as an introduction to most of the simulations in this chapter. We get to use a random number function right away.

- Make a new case table and give it 36 cases. (Right-click on the table, and choose **New Cases...**)
- Give it two attributes, **die1** and **die2**.
- Give **die1** the formula **randomPick(1, 2, 3, 4, 5, 6)**. (You could also have used **randomInteger(1, 6)** or **randomInteger(6)**.)
- Give **die2** the same formula. See the random numbers?
- Make a new attribute, **sum**, and give it a formula that adds **die1** and **die2**.
- Graph **sum**, and make the graph a histogram. Use the graph to answer questions 7 and 8 in the textbook.

You can tell how many rolls are in a histogram bin by pointing at the bin and looking in the status bar at the lower left of the Fathom window. It will show you the exact frequency. This can be useful, especially when there are many cases in one bin.

You can re-roll your 36 pairs of dice instantly by choosing **Rerandomize** from the **Collection** menu, or by typing its shortcut, CONTROL-**Y** (⌘-**Y** on the Mac).

7.1 Discrete and Continuous Random Variables

Example 7.2

❖ Open the file **EG07_02 Tossing Coins.ftm**. It contains a simulation of the coin toss in the example. Press the **Collect More Measures** button (in the collection with the green balls) to get a new set of 160 tosses of four coins.

✐ How do the empirical graphs that you get compare with the theoretical probability histogram in the textbook?

✐ Explain what you see in the **sequence** histogram when you click on one of the bars in the **nHeads** histogram, and vice versa.

This document uses features we haven't talked about. You will learn about measures in Chapter 8. The coin images are controlled in the **Display** panel of the inspector.

Exercises

7.1 ROLL OF THE DIE. You cannot easily sort random numbers in Fathom because they keep rerandomizing. We'll use a different technique.

(b) Make 100 cases and create a new attribute, **die**. Give it an appropriate formula to make it give random integers from 1 to 6.

Make a new summary table. Drag **die** to the left side of the table but *hold down the* SHIFT *key* as you drop **die** onto the table.

✐ What happens if you don't hold down SHIFT? Try it and see.

(c) After you have done 100 rolls five times as suggested, just make 400 more cases. Now you can do 500 rolls all at once.

✐ Is that any different? Why or why not?

7.10 VIOLENCE IN SCHOOLS II. To make the simulation in Fathom, you need to draw numbers from a Normal distribution with mean 0.04 and standard deviation 0.023. It doesn't matter how many you draw.

❖ Make a new table and a new attribute, called **p_hat**. Make 1000 cases.

❖ Give **p_hat** the formula **randomNormal(0.4, 0.023)**. Graph the result.

Now we need to find out how many of those are bigger than 0.45. There are several strategies; again, sorting is not one of them. Here is a really great one:

❖ Make a new slider. Call it **limit**, and set it to 0.45.

❖ Make a new attribute. Call it **subset**, and give it the (Boolean!) formula **p_hat** > **limit**. Graph **subset** in a new graph. You'll get a bar graph.

❖ Click on the bar for **subset** = **true**. What happens in the **p_hat** graph? Why?

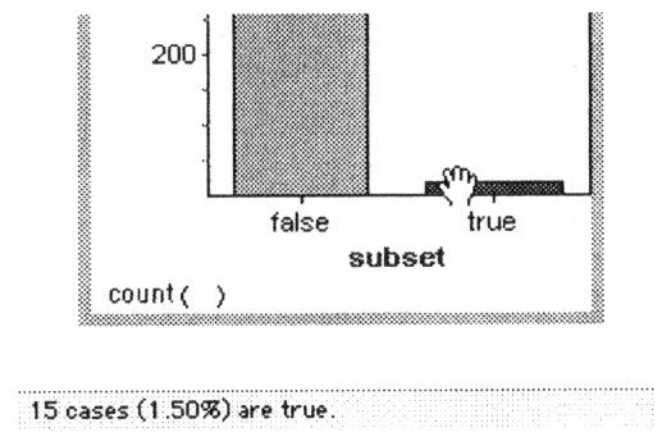

- ❖ To find the number of cases in the subset, point at the bar and read the number from the status bar in the lower-left corner of the screen.
- ❖ Use this setup to answer part (b) of Exercise 7.9. Change the value of **limit** and click on the **false** bar instead of **true**.

  ✎ Why click on the other bar?

  ✎ Compare your results to your answers to Exercise 7.9 parts (a) and (b) in the textbook.

Digression: Why the Normal is wrong, how to fix it, and why it doesn't matter much

Look at the values for **p_hat** (one of ours is 0.365829). You can tell just by looking that they cannot be the results of a poll of 400 people. (How?) The Normal distribution is only an approximation.

To avoid that, use the *binomial* distribution. In Fathom, make a new attribute, **yessers**, and give it the formula **randomBinomial(400, 0.4)**. That tells you how many people answered yes. Use that to compute a new **p_hat** (maybe **b_hat**) and use that to finish the problem. Are the answers that you get very different?

7.13 SIZE OF AMERICAN HOUSEHOLDS, II. Consider this extension:

We downloaded 5000 individuals as Census microdata into Fathom, and looked at the "Family size" attribute in order to compare it to the distribution in the text. The Census assigns individuals living alone a family size of 1. There were 771 of these individuals in our sample, leaving 4229 people with families as defined in the text (i.e., 2 or more people). The distribution of family sizes in this set of 4229 was as follows:

number of persons	2	3	4	5	6	7	8+
frequency in 4229	1256	907	1041	592	245	98	90
probability in 4229	0.30	0.21	0.25	0.14	0.06	0.02	0.02
probability from text	0.42	0.23	0.21	0.09	0.03	0.02	—

Compare this distribution to the one in the textbook (also in the table). You will find it similar overall, but not matching very well. The problem is that in the text, they counted *families*, but here, we counted *individual people*.

- ❖ Using the data in the table, convert the distribution to be "by family" so that you can compare it to the one in the textbook. (Don't use Fathom for this.) Hint: Our sample has 1256 individuals in 2-person families. If they were all living together, how many families would that be?

  ✎ How did you do the conversion? Explain the calculations you made and why.

✐ Explain, clearly and without formulas, what difference it makes in the distribution whether you count people or count families.

✐ This new "converted" distribution is more like the one in the textbook than the first, raw one is. How can you tell it's closer? Be as specific as possible.

✐ Both sets of data—ours and the one in the textbook—were taken from the 2000 Census. Why don't the distributions match exactly?

7.18 THE SUM OF TWO RANDOM DECIMALS. For part (d), use Fathom for the simulation:

❖ Make a table with three attributes, **X1**, **X2**, and **Y**, and 1000 cases. (Why stop at 200? Having 1000 will make calculations easier!)

❖ Give **X1** and **X2** this formula: **random( )**. Make **Y** = **X1** + **X2**.

❖ Make a graph of **Y**. You'll get a dot plot with a triangular shape.

❖ Explore: Make a second graph, this time, a scatter plot of **X1** versus **X2**. Now use your mouse to select points in the **Y** graph.

✐ Explain, without using formulas, why the distribution of **Y** is higher in the middle than on the edges.

Use that graph to answer questions (b) and (c) in the textbook. You can use any of the techniques we have described so far, for example, the one using the **limit** slider and the **subset** attribute we used in Exercise 7.10, above. But here is another one for question (c):

❖ Drag a rectangle that encloses all of the points where **Y** < 0.5, as in the figure.

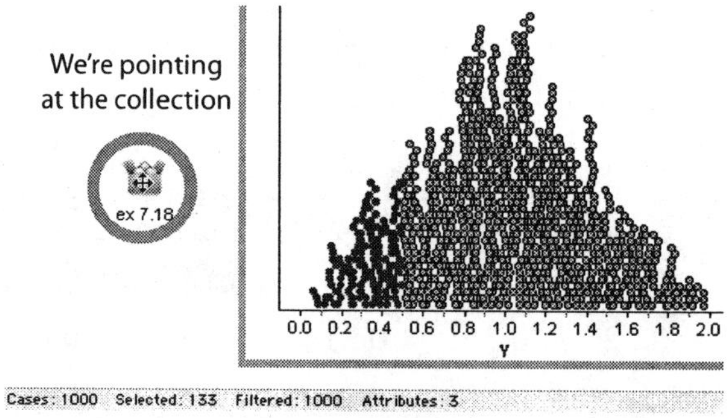

The number of selected points appears in the status bar at the lower left of the Fathom window.

❖ Point at the collection (the box of gold balls).

❖ As long as the mouse is over the collection, the number of selected points appears in the status bar at the bottom. Use that number to help you answer the question.

❖ Rerandomize to repeat the process a few times.

What spread of values do you get? Does that range encompass the theoretical prediction?

Why do the values vary? List at least two good reasons.

Extra! Mercrutio decides to find the distribution of the sum of two random numbers using one attribute, **Z** (instead of three, as we did). He gives **Z** the formula **2 * random( )**. Try it.

This doesn't work. Why not?

How could he have found the distribution using only one attribute? What would its formula be?

7.19 PICTURING A DISTRIBUTION. Continuing from 7.18...

- In the **Y** graph, choose **Histogram** from the menu in the corner of the graph.
- Right-click on the graph and choose **Scale>Density** from the context menu.

Now you'll enter a function so you can see the distribution triangle superimposed on the histogram.

- Right-click again and choose **Plot Function** from the context menu. The formula editor appears.
- Enter the following:

if(Y < 1) TAB **Y** TAB **2 – Y**

The formula editor will look like the illustration.

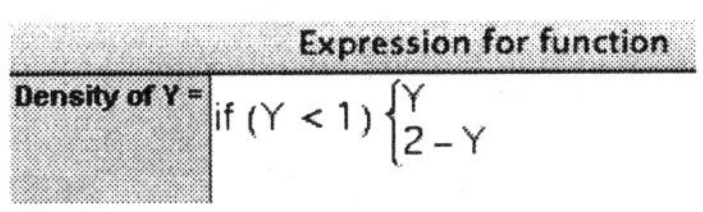

- Press **OK** to make the curve appear on the graph.

How well does this "curve" fit the histogram?

- Using your mouse, draw a rectangle and select all cases where **Y** < 0.5 (as before in Exercise 7.18).

Why is it easier to select the right cases now? (This makes it easier to answer (b) and (c) in 7.18.)

Explain the formula. Especially, explain the **Y** and **2 – Y** to the right of the bracket.

7.21 JOGGERS, II. Do the simulation in Fathom instead of using a calculator. Note that this is similar in structure to Exercise 7.10, above.

- Make a simulation in order to approximate an answer to 7.20(a) in the textbook.

✐ Describe how you made your simulation.

✐ About how many cases do you need in your collection so that every time you rerandomize, you get an answer close to 7.20(a)?

7.2 Means and Variances of Random Variables

The textbook begins this section by showing you how to do *theoretical* calculations. The opening activity 7B is about the set of numbers {1, 1, 2, 3, 5, 8}. You already know how to find the mean and variance of this population using Fathom. You've now learned how to take samples. So see what happens when you take repeated samples of two numbers and compute their means. Your result should be different from the activity, in which you find the mean of all subsets of size two, but the more samples you take, the closer you will get.

Similarly, you can simulate the Tri-State Lottery (Example 7.5).

We return to Fathom's forte, simulation, with Exercise 7.31:

7.31 LAW OF LARGE NUMBERS SIMULATION. Do this simulation with Fathom instead of on your calculator.

- Make a new table with attributes **person**, **height**, **sumSoFar**, and **meanSoFar**. Give the collection 200 cases.
- Give the attributes the following formulas:

person	caseIndex
height	randomNormal(64.5, 2.5)
sumSoFar	height + prev(sumSoFar)
meanSoFar	sumSoFar/person

- Make a graph with **person** on the horizontal axis and **height** on the vertical.
- Duplicate that graph twice (so now you have three identical graphs). Arrange them so you can see them all.
- Put **sumSoFar** on the vertical axis of one new graph, and **meanSoFar** on the vertical axis of the other.
- Rerandomize repeatedly (CONTROL-**Y**; ⌘-**Y** on a Mac) and see how the graphs change.

  ✐ Describe what you see.

  ✐ Explain clearly how the calculation of **meanSoFar** works. Why does its value equal the *cumulative* mean of **height** for all cases up to the one you're looking at?

  ✐ Write a short description of the principle this exercise demonstrates.

> **Graphs versus numbers**
>
> You have been using graphs to see the distributions of random variables. In this section, you will calculate numbers such as means and variances that describe the distributions. It's easy to think of this as becoming *more serious*, that numbers are better than pictures.
>
> Don't be deceived. It's true that means and variances are quantitative, and there is a lot to be said for that. But they never tell the whole story. Whenever you report a mean without the distribution, *you're throwing data away*. We need to do that sometimes; there's too much data to take it all in, and the mean (or the variance, or the median, or whatever) is a useful summary.
>
> But never think that any summary is the whole picture.

7.33 A GAME OF CHANCE. We'll set up the game in Fathom and collect the results by typing them in by hand.

- ❖ Set up a simulation of tossing a coin ten times. This should be easy by now; a good formula for the randomness is **randomPick("H", "T")**. Be sure you have the quotes in there.
- ❖ Make a new table for results. Make new attributes for **maxRunLength** and **outcome**.
- ❖ Enter the result from your simulation: put the longest run of heads or tails in **maxRunLength**, and also record the **outcome**: **+2** if the longest run is less than three; **−1** if the longest run is three or more.
- ❖ Drag a new summary table off the shelf and then drag **outcome** either to the top or the side. The mean outcome will appear.
- ❖ Rerandomize and record the new results. Repeat that many times. Your file might look something like this after two games:

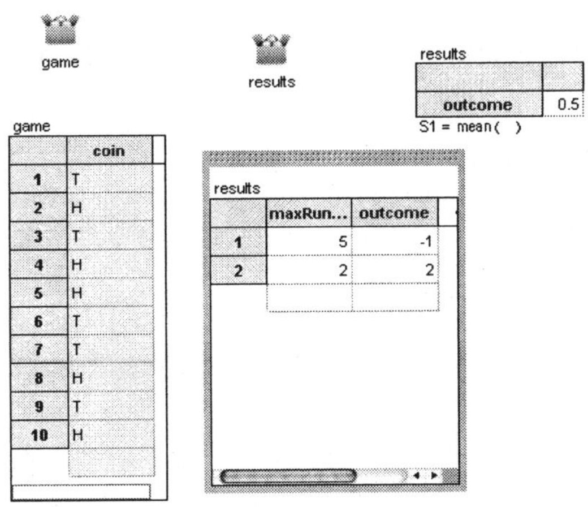

✎ Is it to your advantage to play?

✎ How many games did you need to simulate to feel sure of your result?

LOOKING AHEAD TO MEASURES. You can collect the results automatically, though it requires a new function and collecting measures. Use **max(runLength(coin))** to get the maximum run length. You can find the function **runLength( )** under **Statistical> Transformations** in the formula browser to see how it works. You'll learn about collecting measures in Chapter 8.

Rules for means, an alternative view

The textbook gives you some rules for means. Rule 2, for example, is

$$\mu_{X+Y} = \mu_X + \mu_Y$$

Just in case that isn't clear, here's another way to look at it. If you have two random variables X and Y, you can make a new random variable Z by adding them. That is, if $Z = X + Y$, then

$$\mu_Z = \mu_X + \mu_Y$$

This says the same thing, but it lets you think explicitly about a new variable instead of relying on the expression $X + Y$ down there in the subscript. Also, in Fathom, the easy way to study this rule is to make a new attribute—a whole new column—and look at its mean. The same holds for rule 1, and for the variance rules. The lesson? If making a whole new variable helps you understand, do so.

Simulating the rules for mean and variance

Let's use Fathom to see how well these rules work.

❖ Make a new table with three attributes, **X**, **Y**, and **Z**. Make **X** and **Y** each **random( )**, and make **Z = X + Y**. Give the table some cases.

❖ Drag a summary table from the shelf.

❖ Drag **X** to the top box, then drag **Y** to the *arrow* that points to it, then drag **Z** to the arrow as well. You will see all three means:

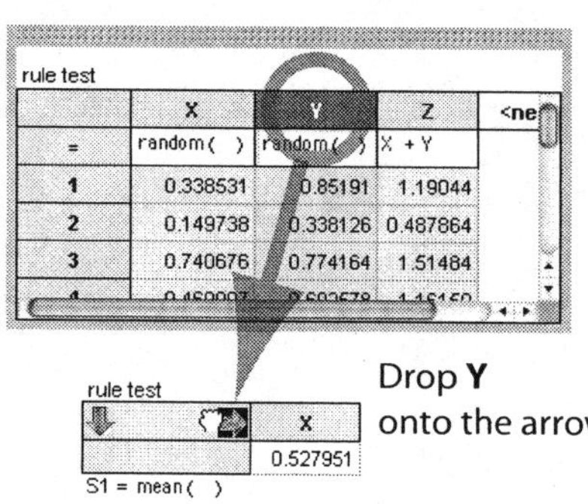

Drop **Y** onto the arrow.

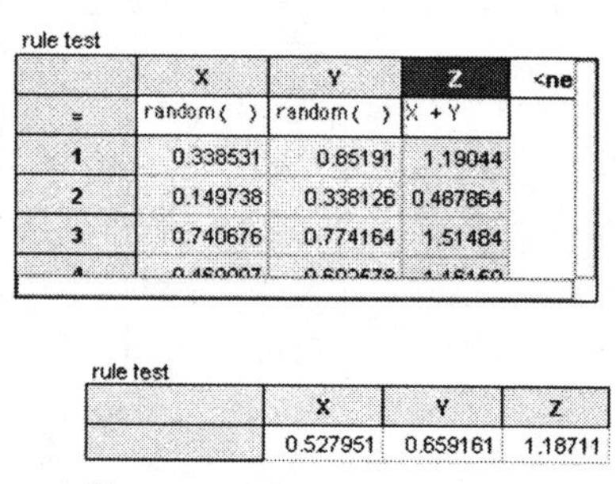

❖ Put the standard deviation and variance in the table as well. (Right-click on the table and choose **Add Formula** from the context menu. Use **StdDev( )** for standard deviation and **Variance( )** for variance.)

✐ Do the means add up? Do they add *exactly*?

✐ Do the variances add? Do they add *exactly*?

❖ Explore. Rerandomize and see what changes. Change the formulas for **X** or **Y** and see if the conclusions still hold. For example, change **Y** to **randomNormal(100)** and see what happens.

✐ Challenge: Why do means add exactly while variances do not?

Example 7.12 SAT scores, extension

What does the SAT data look like? In your Fathom sample documents (use **File>Open Sample Document**), open **SATGPA.ftm**. (It's under **Social Science**, and then under **United States**.) This file contains SAT and GPA data for 1000 college froshpersons during some year in the 1990s.

❖ Using a summary table, compute the mean, standard deviation, and variance for the math and verbal SAT scores. (Use a summary table as you did in *Simulating the rules for mean and variance*, above.)

❖ Make a scatter plot of **verbal** against **math**.

✐ Do these two attributes seem to be related? How can you tell from the graph?

❖ Make a new attribute, **totalSAT**, to be **verbal** + **math**.

❖ Compute **totalSAT**'s mean, standard deviation, and variance.

✐ Check whether Rule 2 for means holds. (It should.)

✐ Check whether Rule 2 for variances holds. (It should not. Describe how you know, even though this rule does not generally hold *exactly* for a sample.)

This demonstrates how important independence is. **Verbal** and **math** are positively correlated, as we see in the graph; they are *not* independent. If you do well on the math SAT, you tend to do well on the verbal.

✐ Context check: How do the scores from the file compare to the scores listed in the textbook? What could explain the differences?

✐ Challenge: If **math** and **verbal** are correlated, then if you know the math score, you know the verbal score better. So if you add them, it's as if there is less variation in the combination. Why, then, is the variance of the sum *more* than the sum of the variances?

Example 7.14 A round of golf, extension

In fact, Tom's and George's golf scores will all be integers. The treatment in the textbook, answering the question, "What is the probability that Tom will win?" neglects this. This matters partly because of the chance that they will tie.

✐ Why is that not a problem when you use the regular Normal distribution?

✐ Will the possibility of a tie increase or decrease Tom's chances to win?

❖ Do a simulation to test your idea and report: How much does it matter that the textbook did not use integers in its solution?

Exercises

7.38 CHECKING INDEPENDENCE, I, EXTENSION. What do the data really look like?

❖ For part (c), go online and see if you can find the data.

The author used "rainfall data Orlando airport" as a search string and found good material buried deep in **www.cityoforlando.net**. This does not have Disney World rainfall data, but the figure shows the daily rainfall for October 2005 at the airport (OIA) plotted against the rainfall at Lake of the Woods, which is not too far away.

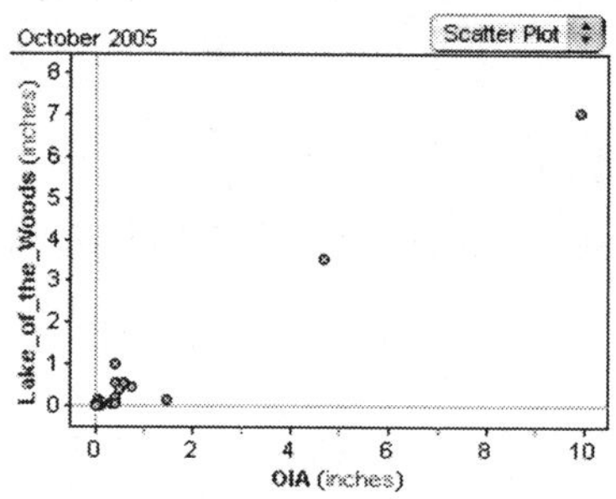

❖ Try to use Rule 2 for variances to show that the two sets of rainfall data are *not* independent.

✐ How did you set that up?

✐ What is your result?

7.43 KIDS AND TOYS. In part (b), the textbook asks you to describe and carry out a simulation of kids with toys, in order to approximate the mean and standard deviation you calculated in part (a).

In Fathom, make a table with a lot of cases—one for each kid—with columns for **nToys** (the number of toys) and **r**, which will be a random number between 0 and 1.

The hard part is simulating the discrete distribution. The picture shows one way to make a formula that works:

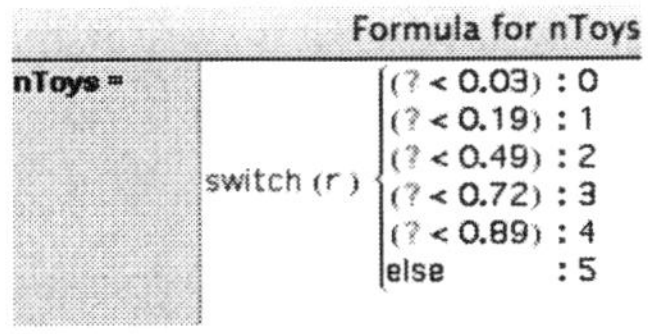

The weird thing here is the **switch** statement. When you first make it, it has only two possibilities. To make a new possible outcome—a new branch—put your cursor in a branch and press OPTION-RETURN (Mac) or CTRL-ENTER (Windows). If it's not clear to you how this statement works, check out Fathom's help system.

✎ Where did the numbers in the **switch** statement come from?

✎ We made a separate column for **r**. What happens if we change the function to have **switch(random( ))**?

7.46 WEIRD DICE AGAIN. In part (c), we can also simulate the situation. The formula for the first weird die is **randomPick(1, 3, 4, 5, 6, 8)**.

❖ Make the simulation and find the distribution of the sum of the two dice.

✎ Also find the mean value of that sum. How does your result compare to the two values you calculated in (c)?

✎ Before, we noticed that the rule for adding means is exact, even when we are simulating. Why doesn't your simulation give exactly the same result as theory?

7.48 A GLASS ACT, II. Simulate 7.47(a) in Fathom.

❖ Use the **switch** technique from exercise 7.43 to set up the probability distribution for temperature.

❖ Simulate many (more than 100) readings, and compute the mean and standard deviation of the result.

✎ Are the mean and standard deviation produced from your simulation close to the values you calculated in 7.47(a)?

7.52 LOSER BUYS THE PIZZA. After you find the result theoretically, simulate it in Fathom. You'll have attributes for Fred and Leona, both with the same formula, **randomNormal(24, 2)**. You'll need another one for the absolute difference in their scores (use vertical bars for absolute value in the formula editor).

✎ How many cases do you feel you needed to get a good answer?

✎ How did you calculate the proportion of trials where the difference was 5 or more points?

- Challenge: ACT scores are integers. Will that make any difference in your result? Without destroying your old simulation, make a new one (perhaps in new columns) with integer scores to test this situation.

  ✐ Describe how you made your integer simulation and how you compared it to the original.

  ✐ Explain without using formulas why it does or does not make a meaningful difference that the scores are integers.

7.62 IT'S A GIRL! We will make the "family" simulation in Fathom; record the results on paper or in Fathom in a new table.

- You need a new table with one attribute, **sex**, and four cases.
- A good formula for **sex** is **randomPick("Boy", "Girl")**.
- Rerandomize in order to get a new family. The size of the family is the position of the first girl, or **4** if they are all boys.

  ✐ Carry out 25 repetitions and then average the results to estimate the expected value.

  ✐ Construct the probability distribution table for the random variable X = number of children.

7.63 SLIM AGAIN. Do all the steps in the textbook, but make the simulation in Fathom.

There are a lot of different ways to construct the simulation. Here are some suggestions:

- Make a table with 50 cases (i.e., 50 rows)—one for each play of the game.
- Since the game has two rounds, you will need at least one column for each round and one column for the outcome.
- Don't worry about ending the game early if you get two heads on the first flips. Go ahead and simulate both rounds of each game—you know you win if you win the first round, no matter what happens in the second.

Here are some ideas about how to simulate the rounds:

- You could use **randomPick("HH", "HT", ...)** (you fill in the dots).
- You could make four columns—one for each coin.
- You will need to use the Boolean buttons in the formula editor, **and**, **or**, and **not**. Do not hesitate to use parentheses to get it right.
- You will need at least one **if( )** statement that figures out whether you won or lost.

8 The Binomial and Geometric Distributions

In this chapter, you will learn about these two important distributions. You'll also learn more about how to set up simulations in Fathom.

In these simulations, you will start using more than one collection at a time. The main new ideas are *collecting measures* and *sampling*.

8.0 Measures and Sampling in Fathom

In the textbook, work that needs measures and samples begins later in the chapter. The chapter begins by using functions that take care of measures and sampling with a single calculation.

In contrast, in this guide, we will begin by stepping you through an example that avoids the new geometric and binomial functions. *This is out of order with the way things are presented in the textbook.* If you find it confusing—just skim it and go on. After this example, we will show you the fancy functions and parallel the textbook exactly. You can refer back to this example when we return to samples and measures in a few pages.

Measures and Corinne (see Examples 8.8, 8.10, and 8.14)

Corinne makes 75% of her free throws in general. She makes 7 out of 12 in one game. We really expect her to make 9 out of 12 (75%). But there's bound to be some variation just due to chance. The question is, is the 7 out of 12 unusual?

If you were making a simulation by hand, you might take these steps:

1. Make a spinner with four sections. Color three of them green for "good" and one maroon for "miss." One spin represents one free throw.
2. Spin the spinner 12 times, recording your results. The number of good free throws is the result for a simulated game. This is one trial. Note that the result is a number: the number of free throws Corinne made.
3. Do that many times, and see how often you get a number that's 7 or less. If it hardly ever happens, maybe Corinne choked. For that game, at least, she was not a 75% shooter. If it happens reasonably often, we can just say her 7/12 performance is due to chance—that is, we don't have any evidence that her 75% free-throw rate changed for the game.

We're going to make our simulation in Fathom using exactly the same steps.

- ❖ In a new document, make a table. Name its one attribute **free** and give it a formula like the one in the illustration. Note how it will give her **good** 75% of the time.
- ❖ Give the table 12 cases. While you're at it, give it a better name than **Collection 1**. We'll call ours **game**, because the 12 free throws represent one game. (Double-click the name to change it.)

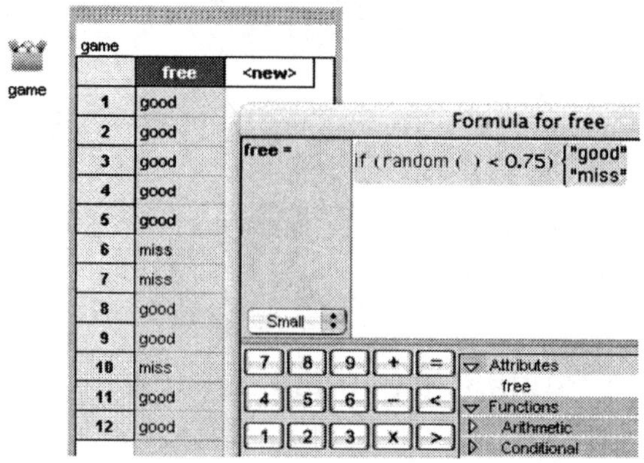

So far everything should look familiar. Each row is one free throw and the whole table represents one game. Now we start the new stuff. We want Fathom to count the number of **good**s in the table.

❖ Double-click the collection—the box of gold balls—to open its *inspector*. The inspector has several tabs—**Cases**, **Measures**, and so forth. The **Cases** tab gives you access to all of your attributes (only **free** in this case) and formulas. Click the **Measures** tab.

❖ Click **<new>** to create a new measure. Name it **Ngood**. Double-click the box in the **formula** column to open the editor. Give it this formula:

count(free = "good"). Read this as, "count of free equals good."

A number should appear—the number of good free throws in your collection. If you rerandomize, that number could change.

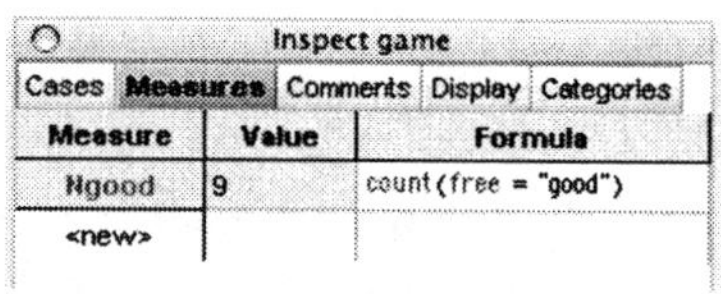

ATTRIBUTES AND MEASURES. A regular attribute—such as **free**—can have a different value for every case—for every free throw. A *measure* like **Ngood** is an attribute for the whole collection. It has one value that summarizes—measures—all of the cases in the collection at once.

The measure, **Ngood**, is useful—it keeps you from having to count the **good**s—but what we really want is to repeat this many times and have Fathom keep track of the numbers. That's next.

❖ Right-click on the collection and choose **Collect Measures** from the context menu.

A new collection appears, called **Measures from game**, and green balls fly to it.

❖ Click once on the new collection to select it. Then drag a new table down from the shelf. You'll see that there are five cases in the new collection. Each case has the number of **good** free throws in one game.

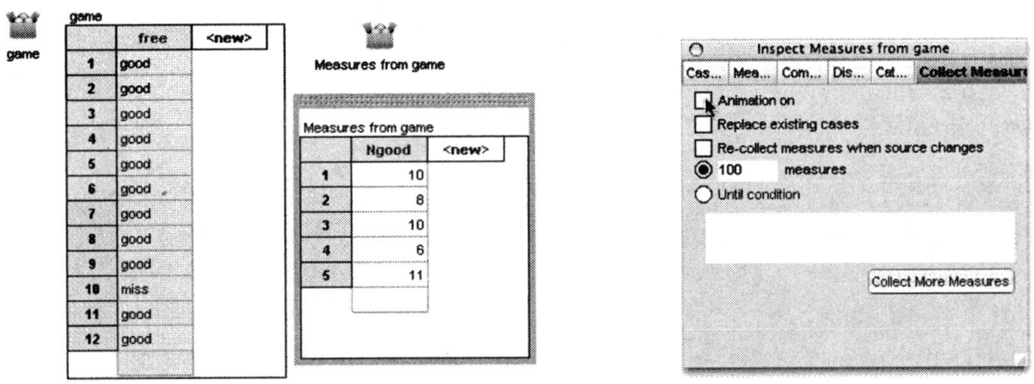

❖ Five games is not enough. Double-click the new collection to open its inspector. You'll see that it opens to its **Measures** tab. There you can control how many measures it collects. Have it collect at least 95 more. Press **Collect More Measures** to make it happen.

You can also control other things such as whether animation is on. In the illustration, we are telling Fathom to turn off animation (this makes it run faster, but the green balls do not fly) and to collect 100 measures.

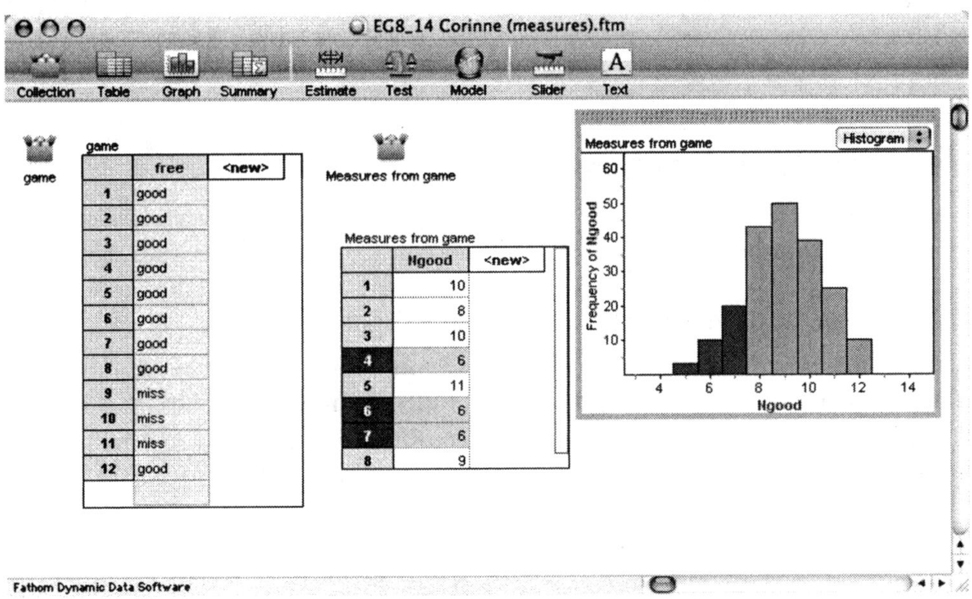

❖ Graph **Ngood**. How unusual is it for Corinne to hit 7 or fewer free throws?

In our example, we have 200 cases in our measures collection. We made a histogram of **Ngood** and selected the bars where **Ngood** is less than or equal to 7. Pointing at the collection and looking at the status bar in the lower left, we see that there are 33 selected cases out of 200—which is 16.5%. We might get a different figure if we did it again.

The graph shows the distribution of the values of the random variable **Ngood**. Notice that it's also about as likely that Corinne will hit 11 or more and still be a 75% shooter. The distribution looks more or less Normal but it's not; it's an empirical example of a *binomial* distribution. It has two parameters: the number of choices n and the probability of a "success" p on each choice. We'll write it like this: $B(12, 0.75)$.

Let's take a moment to compare real life, the manual (spinner) simulation, and the simulation in Fathom:

Real Life	Manual (by hand)	Fathom
Corinne and the ball	The spinner	The formula with **if random( ) < 0.75**
One free throw	One spin	One row in the table
One game	12 spins	the **game** table and collection
Counting 7 good free throws	Counting the green spins and recording the result	The **Ngood** measure and its formula
Lots of games	Lots of spinning, counting, and recording	Collecting measures

COLLECTING MEASURES IS POWERFUL. You can collect measures to solve a wide variety of problems in statistics. It may seem like a lot of work—opening inspectors, writing formulas, dragging collections and tables around—but it's an investment that pays off.

8.1 The Binomial Distributions

Now we will return to the textbook. For the moment, we'll set aside what you just learned about measures and present Fathom's built-in functions for binomial situations.

Example 8.7 Inspecting switches: Calculating a binomial probability

The textbook tells you how to find values for the binomial probability density function (pdf). Fathom has functions to handle this. The function **binomialProbability( )** makes the relevant calculation. If the probability of success for one event is 0.1, and you watch 10 events, the overall probability that exactly one of the 10 is a success is:

binomialProbability(1, 10, 0.1)

How do you see the value, though? Fathom is built to calculate statistics about whole collections; how do you calculate one thing? One way is to use a *computed slider*.

❖ Drag a slider off the shelf. Double-click the slider axis to bring up its inspector.

❖ Double-click the box in the **Formula** column to open the formula editor.

❖ Enter the formula you want, e.g., **binomialProbability(1, 10, 0.1)**, then click **OK**.

Presto! The inspector and the slider show the result. (You can also use the inspector to change the bounds of the slider; the range [0, 1] is appropriate here because the number is a probability.) Also, because the slider is *computed*, you can no longer drag the pointer.

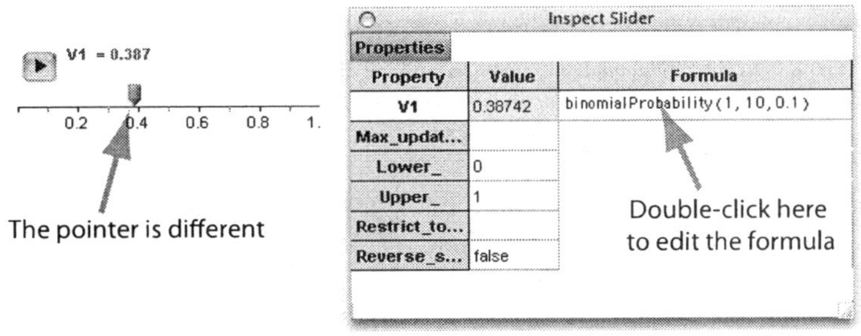

A computed slider and its inspector. Notice how the pointer looks different from that of a regular slider.

Fathom's order of arguments is different from the calculator's. You can think of this function in general as **binomialProbability(x, n, p)**, where **x** is the number of successes in **n** trials, each with a probability **p** of success.

You can find information on any Fathom functions in the formula browser—the box at the lower right of the formula editor. This one is under **Functions>Distributions>Binomial.**

To answer the actual question in the textbook, you need to find the probability for 0 or 1 defective switches. Here are two strategies for finishing the problem:

1. Double-click the formula box in the inspector and add **binomialProbability(0, 10, 0.1)** to the formula that's already there.
2. Make *another* slider for **binomialProbability(0, 10, 0.1)**, and add them in a third!

Note that these are *exact* theoretical probabilities. In our simulation with Corinne, where we collected measures, we got an approximation.

WHY USE MEASURES AT ALL? If we have these perfect functions, why would we ever go to the trouble of collecting measures? Two reasons: first, it's easier to tell what is really going on; the function, for all its speed and power, is a "black box." Second, there are situations where we don't have a function to give us an exact theoretical answer. Then we have to simulate.

Example 8.8 Corinne's free throws: Calculating binomial probabilities

In this example, we again calculate exact probabilities. But we want to add all of them from **binomialProbability(0, 12, 0.75)** to **binomialProbability(7, 12, 0.75)**. That's a pain with sliders. So we'll use other Fathom capabilities we already know:

- Make a table and give it two attributes, **X** and **p**.
- Type in values for **X** sequentially from **0** to **7** (so you have eight cases).

❖ Give **p** the formula **binomialProbability(x, 12, 0.75)**. Aha! The probability table fills in!

❖ Now we have to add them up. Make a summary table, and drag **p** to the top or the side (it doesn't matter which). The mean of **p** appears.

❖ The mean is irrelevant, so double-click the formula at the bottom and edit it to be **sum( )** instead of **mean( )**. Done! Save your work.[1]

EG8_06 Corinne free throws

EG8_06 Corinne free throws

p	0.15764368

S1 = sum( )

EG8_06 Corinne free throws

	X	p	<new>
=		binomialProbability(x , 12, 0.75)	
1	0	5.96046e-08	
2	1	2.14577e-06	
3	2	3.54052e-05	
4	3	0.000354052	
5	4	0.00238985	
6	5	0.0114713	
7	6	0.0401495	
8	7	0.103241	

Example 8.10 Is Corinne in a slump? Binomial probabilities by formula

Instead of taking a sum, as we did in Example 8.6, we could use a *cumulative distribution function.*

❖ Use the file you saved in Example 8.6 and make a new attribute, called **F**, and give it the formula **binomialCumulative(x, 12, 0.75)**. Notice that the arguments are in the same order as for **binomialProbability**. Each row of the table now has the cumulative sum of all previous rows—and the entry for 7 is the sum we calculated before.

EG8_08 Corinne redux

	X	p	F
=		binomialProbability(x , 12, 0.75)	binomialCumulative(X , 12, 0.75)
1	0	5.96046e-08	5.96046e-08
2	1	2.14577e-06	2.20537e-06
3	2	3.54052e-05	3.76105e-05
4	3	0.000354052	0.000391662
5	4	0.00238985	0.00278151
6	5	0.0114713	0.0142528
7	6	0.0401495	0.0544022
8	7	0.103241	0.157644

You can use this as a probability table. You can also graph the results. Just put **X** on the horizontal axis and **P** or **F** on the vertical (If you want them both, drop the second one on the plus sign that appears).

If you want to connect the dots, choose **Line Scatter Plot** from the menu in the corner of the graph. Some people think that's inappropriate (see page 49 in this guide for why); if you have Fathom v2.1 or later, you can give the points horizontal error bars (look in the *graph's* inspector); each should extend ± 0.5.

1 Do you like how the formula is in the table? Right-click the table; choose **Show Formulas**. Double-click to edit.

Exercises

You can use Fathom to help with most of the exercises (unless your instructor forbids you, of course!). When you have to use the binomial probability formula, you'll need factorials. In Fathom, just use an exclamation point.

COMPLEMENTARY PROBABILITIES. Don't forget that you can solve a lot of probability problems by finding the probability of something *not* happening, and subtracting that from one.

8.13 INHERITING BLOOD TYPE. Use the same technique we just learned in Examples 8.8 and 8.10. You need a table with three columns: **X**, the number of children with type O blood; **P**, the probability; and **F**, the cumulative function. The names of the variables don't matter as long as they make sense.

Instead of making the histograms, use a scatter plot as in Example 8.8.

8.14 GUESSING ON A TRUE-FALSE QUIZ. A good Fathom technique here is to make a table with three rows (three cases). One attribute is **passingScore**, with values 25, 30, and 32. You have to figure out what other attributes and formulas you need.

The Normal Approximation to the Binomial Distribution

You can use the same techniques we used in Chapter 7 to solve problems using the Normal approximation. But how good is that approximation? You can get a feel for that by playing with the file **08 Normal v Binomial.ftm**. You can drag **n** and **p** and see both distributions. Notice how **mu** and **sigma** change.

✐ *Why* do **mu** and **sigma** change? How did we get that to work?

More exercises

8.25 CORINNE'S FREE THROWS AGAIN! Make a probability table for Corinne's free throws.

- ❖ Make a table with three attributes, **X**, **P**, and **F**. Put the number of free throws made out of 12 into **X** (so you will have 13 cases, from 0 to 12). Put the binomial probability in **P**, and the cumulative probability in **F**.
- ❖ Make a new column, **cumP**, and give it the formula **P + prev(cumP)**.

  ✐ Is **cumP** the same as **F**? Explain why. (What does **prev** do?)

8.26 SIMULATING DEFECTIVE SWITCHES.

- ❖ Set up a table with 100 cases and one attribute, **N_bad**, with random values that give the number of bad switches out of 10, when each has a probability of 0.1 to be bad. (The formula is **N_bad = randomBinomial(10, 0.1)**).
- ❖ Make another attribute, **P_bad** that gives the *proportion* of bad switches. It will be based on **N_bad**.

- Graph **P_bad**.
- Find the average *proportion* of bad switches in your samples.
- Change the simulation to use samples of size 25, and then samples of size 50.
  - ✎ How did the average proportion change when you changed the sample size?
  - ✎ How did the graph change as you changed the sample size?
  - ✎ In reality, the quality engineer gets only one sample. Based on your answers to the previous two questions, why would she want to sample 50 when the average proportion is about the same as when she samples 10?

8.27 EGGS AND SALMONELLA. In part (b) you need to construct a simulation. Make 1000 cases and use **randomBinomial( )**. You will have to figure out what goes inside the parentheses and how to count up the results.

8.28 STUDENT INDEBTEDNESS. We want to know the probability that more than 12 students in a sample of 30 have four or more credit cards. The probability for each student is 0.33.

You could do this just as in the previous problem (make 1000 cases, use **randomBinomial**, and so forth) but let's use measures, just to recall how we did Corinne's free throws back at the top of the chapter.

- Make table with one attribute, named **fourOrMore**. Give the table 30 rows.
- Give **fourOrMore** a suitable formula. Try **random( ) < 0.33**.
- Rename the collection **students**.
- Give **students** a measure, called **bigSpenders**, with a formula that counts how many of the 30 students have four or more credit cards.
- Collect measures—say 1000 times, with animation off—to build up the distribution of **bigSpenders** in the **Measures from students** collection.
- Graph that distribution and answer the question.
  - ✎ Why didn't we need an **if( )** statement on the formula for **fourOrMore**?

8.30 DRAWING POKER CHIPS. In part (c) you need to do a simulation. For this one it really helps to use measures, because there are so many things to keep track of. Your first collection (maybe called **chip sample**) will be the sample of 25 chips. You'll probably need an attribute such as **color**, which randomly has the values **"blue"**, **"white"**, or **"red"** in the right proportions. (For an idea how to do this, see the treatment of Exercise 7.43.)

Then create some measures in your **chip sample** collection. You will need *at least* the number of blue chips so you can answer (d) and (e).

8.37 SIMULATING GRADUATION. Use either the **randomBinomial( )** technique or the measures technique for this simulation.

Binomial Functions in Fathom

In this section, we have used three functions that involve the binomial distribution. Each involves looking at **N** events, each of which has a probability **p** of success.

randomBinomial(N, p)	a random number of successes from that distribution
binomialProbability(x, N, p)	the probability of **x** successes in **N** draws
binomialCumulative(x, N, p)	the probability of **x** *or fewer* successes

8.2 The Geometric Distributions

In the geometric[2] setting, the random variable is how many trials it takes for something to happen (a success) instead of the number of successes in a fixed number of trials. Sometimes people call the geometric distribution the *wait-time* distribution.

Geometric Distribution Functions in Fathom

In this section, you will use three functions to help you solve geometric-distribution problems. The functions depend only on the probability of success **p**.

randomGeometric(p)	a random number of trials required to get a success
geometricProbability(x, p)	the probability that it takes **x** trials for a success
geometricCumulative(x, p)	the probability that it takes **x** *or fewer* trials

These functions are analogous to the ones for binomial situations, above. And as with the binomial distribution, solving geometric problems using functions is fast, but doesn't really show you what is happening. So we will begin by showing you how to study a geometric situation using the measures machinery, which is slower but clearer and more general.

As you will see, we will use one more collection to make this work. We'll need a source collection, a sample collection, and a measures collection.

Example 8.15 Roll a die, extended (the "measures" version)

We will roll a die until a "3" appears. How many rolls will it take?

- ❖ First create your source table. Give it one attribute, **pips**, and give pips six values: **1**, **2**, **3**, **4**, **5**, and **6**. Name the collection **die**. This is the die you're going to roll.
- ❖ Right-click on the collection (the box named **die**) and choose **Sample Cases** from the context menu. A new collection appears (**Sample of die**) and some blue balls fly to it.
- ❖ Make a table of the sample collection so you can see what's in it.

---

2 Here, *geometric* has the same meaning as in "geometric sequence," where each term has the same ratio to its predecessor.

You will see that it has ten samples from the 6 numbers in **die**. Evidently it is sampling *with replacement.* Fathom samples ten cases by default. ("With replacement" is good in this case. Why?)

❖ Double-click the sample collection to open its inspector. It opens to the **Sample** tab. There you can tell Fathom to ignore the fixed number and keep sampling *until some condition occurs*, like this:

❖ Click on the **Until condition** button and enter **pips = 3** into the formula editor. Press **OK** to close it, then press **Sample More Cases**. You should see something like the illustration.

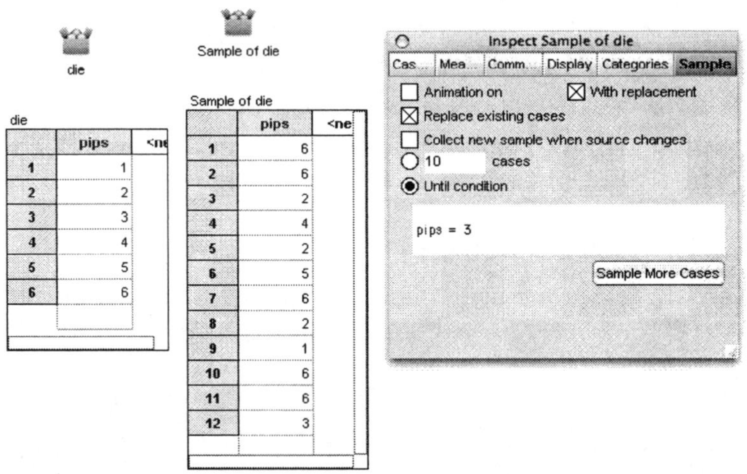

Now this is a geometric situation. Every time you press **Sample More Cases**, Fathom rolls the die until you get a three. Now we want Fathom to count up how many cases there are (that's our random variable; it will be a *measure*).

❖ In the inspector for the **Sample of die** collection, click the **Measures** tab (It probably says **Mea...**). This should look familiar now.

❖ Make a new measure, called **rolls**. Give it the formula **count( )**. The number of cases should appear in the **Value** column. Try it out and see how that number changes.

How not to switch back and forth. It's a pain to click the **Sample** tab, then click **Sample More Cases**, then switch back to the **Measures** panel to see the result. So drag the edge of the **Sample of die** collection a little to open it up. A **Sample More Cases** button appears. That way you can stay in the **Measures** panel and still have access to the button.

Now we want to collect those numbers into a distribution. We'll use **Collect Measures**.

❖ Right-click on the **Sample of die** collection and choose **Collect Measures** from the context menu. Yet another collection appears, this time with green balls, called **Measures from Sample of die**. Make a table to see its contents; you'll see five values for **rolls**. (Fathom collects five measures by default.)

❖ Double-click the new (blue-ball) collection to open its inspector. In its **Collect Measures** panel, change the number of measures so you collect a lot of them.

❖ Collect the measures and graph the result. Turn off animation if things are too slow. Your (crowded) window might look something like the next figure.

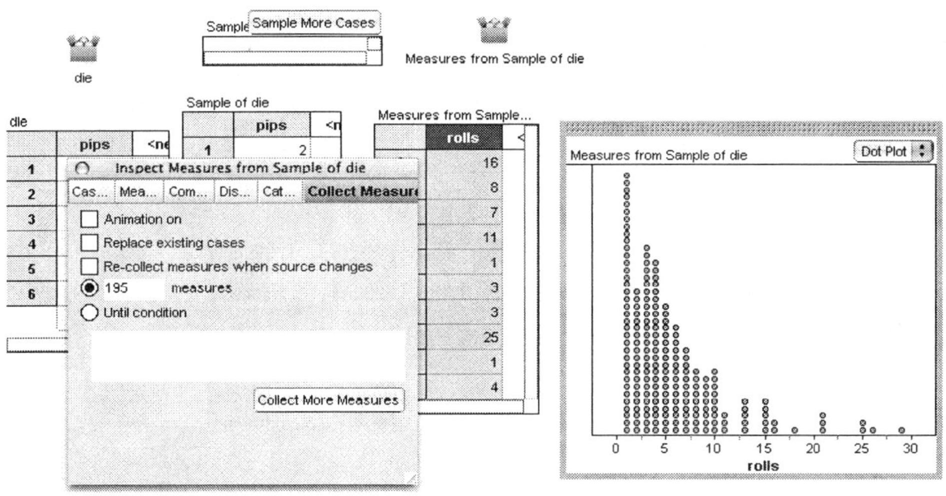

SCREEN SPACE ADVICE. We have three collections, three tables, a graph, and an inspector. In other situations, you may want even more equipment littering your screen. How can you stay tidy? First, shrink tables so that they take up less space. Do you *need* to see that the **die** has numbers 1–6? No. Second, delete or *iconify* objects you don't need. If you drag the corner of an object so that it gets as small as possible, it turns into an icon. Also, *collections* hold data; *tables* do not. If you delete a table, you can just make it again.

❖ Save the file you made.

Refer back to this section when you need to remember how to make this three-collection system. Schematically, it looks like this:

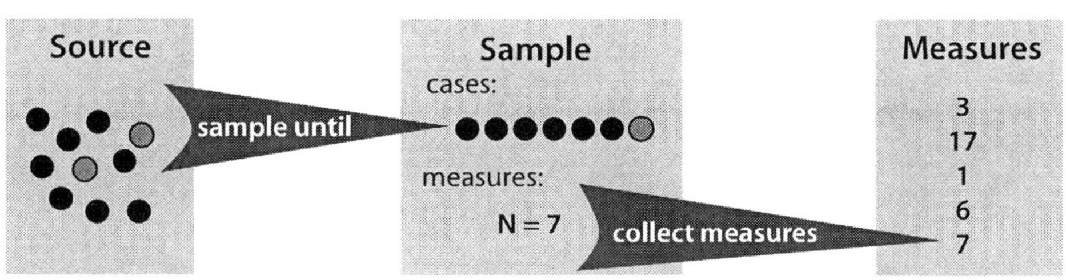

By changing the source and the **until** condition, you can change the problem your simulation solves. Think: what would you change in order to set up the following situations?

- Flip a coin until you get heads.
- Sample parts until you get a defective one.
- Roll dice until you have gotten your *second* **3**.

The Mean and Variance of a Geometric Random Variable: Fathom investigation

Using the data you collected in your measures collection, find the mean and variance of the number of rolls it takes to get a 3. Intuition may tell you that since there are six faces on a die, it should take an average of six rolls to get a 3. Does that seem to be correct?

Connection question: You can approximate the binomial $B(n, p)$ distribution with a Normal distribution under certain conditions. Under what conditions, if any, can you approximate the *geometric* distribution with the Normal? Explain, and illustrate your answer with a Fathom simulation.

Example 8.17 Roll a die: Probability distribution table

Let's make a table that shows the theoretical probabilities of getting different values in a geometric situation. We'll use formulas here because sampling and collecting measures gives us only an empirical estimate of those probabilities.

Let's keep looking at the situation as above, in which we roll a die until we get a 3. In this case, the probability of success is ⅙.

- Make a table with four attributes. Name them **X**, **P**, **F**, and **Pcalc**.
- Type the values 1 through 10 into **X**, so you have 10 cases.
- Give **P** the formula **geometricProbability(X, 1/6)**. This is the probability that you will get the first 3 on the **X**th roll.
- Give **F** the formula **geometricCumulative(X, 1/6)**. This is the probability that you will get the first 3 on the **X**th roll *or before*.
- Make graphs of these two distributions (the regular and the cumulative) to see what they look like. See Example 8.8 for ideas of how to plot them if you have trouble.

Now we want to see if the probability distribution in **P** really follows the theoretical structure described in the textbook. The first probability is correct—the chance of getting a 3 in one roll is ⅙, or about 0.16667. But each subsequent entry in the table should be ⅚ of the previous one. Let's put those values in **Pcalc** and compare them to **P**.

❖ Give **Pcalc** a formula like the one in the illustration.

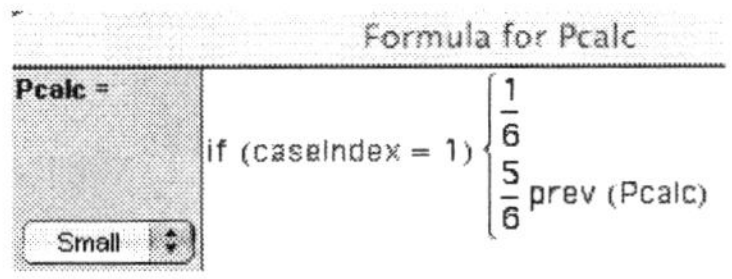

✐ How do **Pcalc** and **P** compare?

✐ Why does the formula for **Pcalc** work? Explain it.

❖ Extension: Make a more general-purpose tool. Drag a slider off the shelf and call it **eventP**. Give it a value less than 1, for example, 0.1.

❖ Edit the formulas for **P** and **F**, and substitute **eventP** in place of ⅙. What would you substitute in the formula for **Pcalc**?

❖ Save this file!

Now the **eventP** slider controls the probability of the individual events—and the table and all of the graphs update accordingly.

✐ What happens when **eventP** is not between 0 and 1? Why?

Exercises

8.50 Game of chance. Use Fathom to make the simulation in step (h). You could use Fathom's **randomInteger( )** function, but it may be clearer to you what is happening if you modify the source-sample-measure technique. Here are hints for a "source-*measure*-measure" strategy:

❖ The source should have three cases, each of which is a random coin toss. Call the attribute **coin**. In this case, it might help to code heads as **1** and tails as **0**.

❖ That collection has a measure, maybe called **winner**, that is **true** whenever there is a winner in the game. What values for **sum(coin)** indicate a winner?

❖ Instead of sampling, collect measures until you get a winner. Calculate a measure in *that* collection—the number of tosses you needed to get a winner—which is the number of cases in that measures collection.

❖ Now collect measures from that measures collection to build up your distribution and answer the questions in the textbook.

Strange names. The last collection will be named something like **Measures from Measures from toss**. Weird, huh? But it makes sense.

8.51 Drawing Marbles, I. This is a straightforward activity; make tables just as you did in Exercise 8.17. But be careful about the formulas. What is the probability of drawing a red marble on the first draw? What does that correspond to in the roll-a-3 scenario in 8.17?

8.52 DRAWING MARBLES II. This is not like the previous one. It needs the source-sample-measure technique. The "until" condition is **count(color = "R") > 1**, or the equivalent.

✐ How did you set up your source collection?

✐ Are any requirements for a geometric setting violated here? If so, which one(s)?

8.54 IT'S A BOY! You could do this either with formulas or with source-sample-measure.

✐ How did you figure out the expected number of girls?

8.57 FAMILY PLANNING, III. If you do the simulation in Fathom, you must use the source-sample-measure technique.

✐ What did you use for your source?

✐ What did you use for an **until** condition in your sample collection?

✐ What measure(s) did you collect?

✐ What do you get for the chance of having a girl at all? Compare that to the result in 8.55(f).

✐ How do your simulated results compare with the theoretical expected value of 1.8 children?

8.58 MAKING THE CONNECTION. Make the table vertically, of course, then just make a scatter plot in the normal way. Follow the steps in the textbook.

✐ What formula did you use for **Y**?

9 Sampling Distributions

The chapter begins with a case study about batteries. Let's make a simulation in Fathom to shed some light on the batteries and on the whole topic of the chapter.

❖ Open the file **CS09 Building Better Batts.ftm**. It has a collection that contains the battery lifetimes (**life**) from page 561 of the textbook.

❖ Make a dot plot of the data, and use **Plot Value** to plot the **mean( )**.

The mean lifetime of the sample is 16.7 hours. Batteries from this factory are supposed to have a mean lifetime of 17 hours with a standard deviation of 0.8. Does the difference indicate that something is wrong, or should you expect a mean as low as 16.7 hours pretty often, just by chance?

Let's simulate data from the Normal distribution—in fact, from *N*(17, 0.8)—and compare it to our test data.

❖ Make a new attribute called **sim** for simulated. Use **randomNormal(17, 0.8) hours** for its formula. The column fills with the simulated lifetimes of batteries.

❖ We want to graph the simulated data next to the real data. Drop the label **sim** onto the same axis where **life** is—but be sure to drop it onto the plus-sign so that you add it to the graph instead of replacing what's there. A **Plot Value** line of the mean appears as well. It should look something like the illustration.

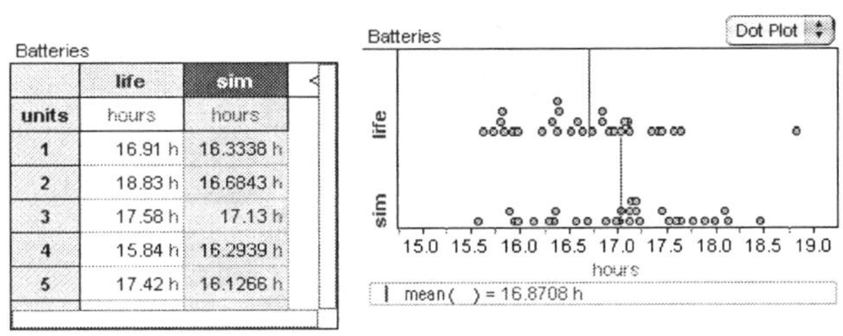

❖ Rerandomize repeatedly, and see how often the mean of our simulated sample is as low or lower than the mean of our test sample.

If we collected all of those sample means, we would have a *sampling distribution of the mean*. We can (and will) do that using the **Collect Measures** mechanism we learned about in Chapter 8.

Activity 9A Young women's heights

This activity uses familiar techniques. Here are instructions for how to do the simulation part in Fathom; use your textbook for the overall instructions.

1. Simulate the heights of 100 randomly selected young women as follows:

- ❖ Make a new collection called **young women** with one attribute, **height**, and 100 cases.

- ❖ Give it this formula: **randomNormal(64.5, 2.5) inches.**

2. Display your data and look at the distribution.

   - ❖ Plot a histogram of **height**. Rerandomize at will.

     ✐ Describe the approximate shape of your histogram. Is it fairly symmetric or clearly skewed?

3. Study how many heights are within 3, 2, or 1 standard deviations from the mean.

   - ❖ Double-click the graph to bring up its inspector. Set the **binAlignmentPosition** to **64.5** and the **binWidth** to **2.5**, as shown.[1] You may need to rescale the vertical axis to see the tops of the bars.

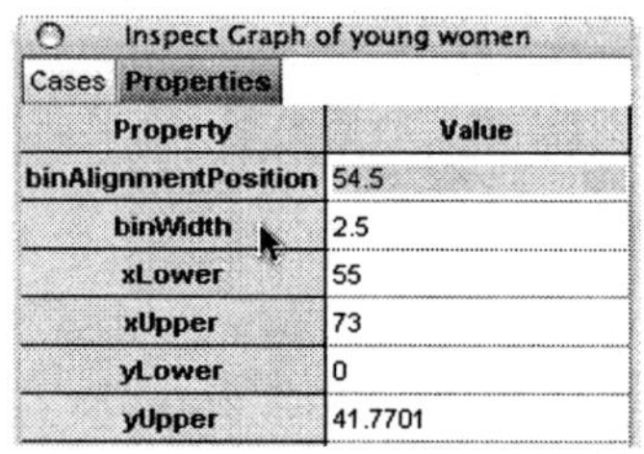

   - ❖ Now the graph is set so that each bin is a standard deviation wide. Select the two middle bins. When you point at the collection, the number of selected cases appears in the status bar at the bottom of the window, as in the illustration below.

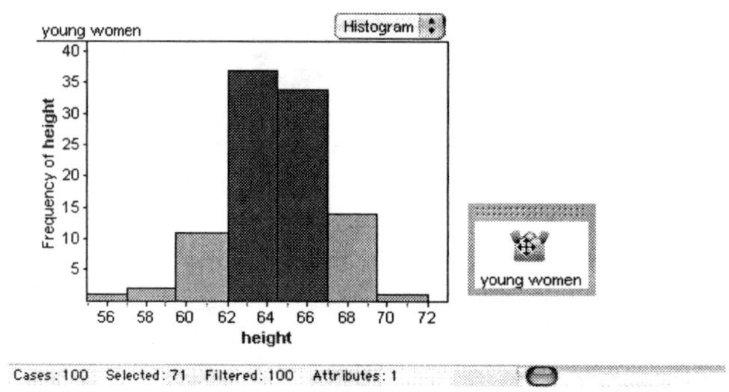

From the numbers in the status bar, we can see that 71% of our cases were within one standard deviation of the mean.

   - ❖ Do the same for two and three standard deviations.

     ✐ How do your figures compare with what you might expect based on your knowledge of the Normal distribution?

1 The value in the **binAlignmentPosition** box may change to that of another bin boundary. That's normal in Fathom 2.1 and earlier.

4. Get numerical statistics.

   ❖ Use a summary table to get the mean, median, and standard deviation. Proceed as in the textbook.

     ✐ Compare the mean and standard deviation of your sample with the mean ($\mu = 64.5$) and standard deviation ($\sigma = 2.5$) of the whole population.

     ✐ How do the mean and median of your 100 **height**s compare? Remember that the more symmetric the distribution, the closer the mean and median are.

5. Make a box plot.

   ❖ Right-click on your histogram and duplicate it: Choose **Duplicate Graph** from the context menu.

   ❖ Change the duplicate graph to a box plot and move it so it does not overlap. (This technique ensures that the scales of the graphs will be the same.)

     ✐ Answer the questions under step 5 in the textbook.

   ❖ **Save** your file for the extension that follows.

6. When you repeat steps 1 to 5, rerandomize in order to draw a new sample.

The rest of the activity follows the textbook exactly.

Extension: Collecting measures

If you do not have a class handy to make the other sticky notes and post them, here are the steps for collecting measures yourself. You may remember all of this from Chapter 8, but it bears repeating as you will do this a lot in this chapter.

Our goal is to collect the mean **heights** of many samples of 100 young women. We begin with our collection of **young women** with 100 cases and one attribute, **height**, which has a random Normal formula. Above, we calculated the mean height in a summary table. Here, we must do it in a *measure*.

❖ Double-click the collection; or right-click it and choose **Inspect Collection** from the context menu. The inspector appears; go to the **Measures** panel.

❖ There, create a new measure called **heightBar** (or some other sensible name). Double-click in its formula column and give it the formula **mean(height)**. The mean appears.

NOTE. This time you really need **height** inside the parentheses. You would use **mean( )** with **Plot Value** or in a summary table, because you want the mean of whatever you put there. For a measure, however, be specific: **mean(height)**.

❖ Right-click on the collection and choose **Collect Measures**. A new collection appears named **Measures from young women**. Green balls fly to it.

- The measures collection gets five values of **heightBar** from five fresh samples of 100 young women. Make a graph of them.
- Inspect that new collection and change its **Collect Measures** panel to collect many more than the five measures it collects by default.
- Study the distribution of **heightBar** just as you would have studied the stickies in the textbook. For example, find its mean and standard deviation—and compare that standard deviation with $\sigma/\sqrt{100}$.

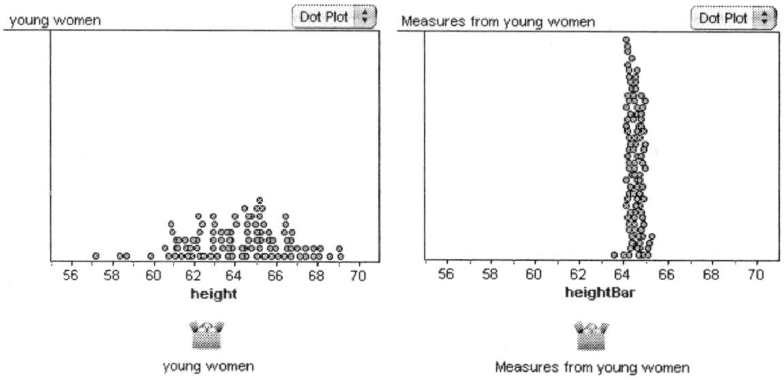

Distributions of **height** and **heightBar**, produced as described. We have forced the axis scales to be the same using *axis links*. You can look them up in Fathom help.

9.1 Sampling Distributions

The section begins describing the difference between a parameter and a statistic. In Fathom, a *parameter* is usually a constant in a formula or, better yet, a slider value. In our formula for young women's heights, **randomNormal(64.5, 2.5)**, the **64.5** (the mean height) was a parameter—a property of the population—as was the **2.5** (its standard deviation).

Statistics are different from parameters because you *compute* them from real or simulated data; they are not inherent properties of the universe we're investigating. You can compute statistics in Fathom in computed sliders, in summary tables, or as *measures*. We will concentrate on the latter here.

Finally, a *sampling distribution* is a distribution of *statistics*. It could be a distribution of *any* statistic—anything you can calculate, such as the mean, the median, the range, the minimum value, or the proportion of cases that fit some criterion. In this chapter we concentrate on the mean and the proportion. And since we want a distribution of these things, in Fathom we'll represent them as measures so we can collect them to make that distribution.

We will often use shortcuts to save us the trouble of setting up measures. Use them, but do so wisely; always remember how to use measures if you have to.

Exercises

9.3 Murphy's Law and tumbling toast. Instead of tossing a coin, we'll do this simulation with measures in Fathom. In the following problem, we'll use the shortcut.

- ❖ Make a table with 20 cases and one attribute, **coin**. Give it the formula **randomPick("H", "T")**. Now we're simulating 20 coins.
- ❖ We want to record the proportion of heads. So make a measure named **pHeads**, and give it a suitable formula such as **count(coin = "H")/20**. Verify that the formula works.
- ❖ **Collect Measures** and make a histogram of the results. Get at least 100 samples.
  - ✏ Is the center close to 0.5? Is the shape approximately Normal?
  - ✏ How much sampling variability is present? That is, how much do your values of **pHeads** from samples of size 20 differ from the actual population proportion, $p = 0.5$?
  - ✏ Get more values of **pHeads** by collecting more measures. How does that change the distribution?
  - ✏ In this simulation, how would you change the number of coins that were tossed? Increase the number of coins to 80 or more. (This requires changes to *two* formulas.)
  - ✏ Collect a new set of measures. How does the distribution of proportions change?

9.4 More tumbling toast. Now we'll see how to do the same thing in Fathom without collecting measures. This is a great shortcut, but be sure to remember the long route.

- ❖ Make a table with 100 cases and one attribute, **nHeads**. Give it the formula **randomBinomial(20, 0.5)**. Now we're simulating 100 samples of 20 coins. Each value is the number of heads among 20 tosses.
- ❖ Make a new attribute, **pHeads**, with formula **nHeads/20**. That's the proportion. Graph it. Rerandomize to get a new set of 100 samples.
  - ✏ Is the center close to 0.5? Is the shape approximately Normal?
  - ✏ How much sampling variability is present? That is, how much do your values of **pHeads** from samples of size 20 differ from the actual population proportion, $p = 0.5$?
  - ✏ Get more values of **pHeads** by adding more cases to the table. How does that change the distribution?
  - ✏ In this simulation, how would you change the number of coins that were tossed? (It will still use only two formulas.)

- ❖ Make the change to n = 80+. How does the distribution of proportions change?
- ❖ **Save** your file for the extension to come.

9.4B GENERALIZING COIN FLIP DISTRIBUTIONS. Let's extend Exercise 9.4. In the formula for **nHeads**, the numbers **20** and **0.5** are *parameters*. Let's represent them by sliders.

- ❖ In your file from Exercise 9.4, make two sliders; call them **n** and **p**. Set **p** equal to **0.5**.
- ❖ Double-click the **n** slider (on the axis, not on the **n**) to open its inspector. Enter the number **1** into the box labeled **Restrict_to_multiples_of**. (It may just look like **Restr...**). This makes **n** an integer-only slider. Set **n** to **20**.

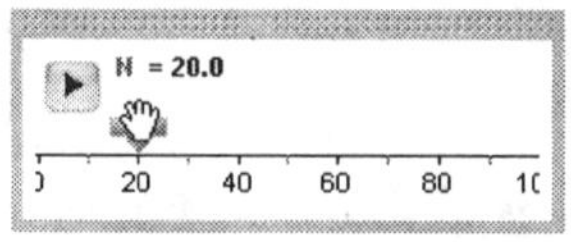

Use the slider's inspector to restrict it to integers.

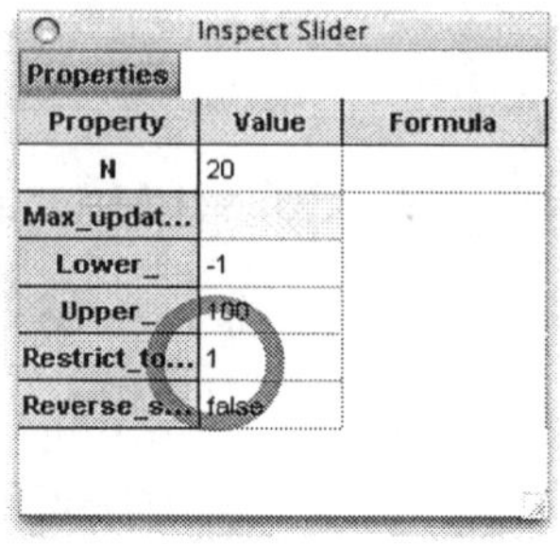

- ❖ Change the formula for **nHeads** to **randomBinomial(n, p)**. Change your formula for **pHeads** to **nHeads/n**. And for generality, change the name of **pHeads** to **pHat** and **nHeads** to **successes**.
- ❖ Make sure your graph is a histogram.
- ❖ Rerandomize, and play with the values of **n** and **p**.
  - ✎ What do you change to make the distribution of **pHat** thinner?
- ❖ Make a graph of **successes** as well, and keep playing.
  - ✎ Why do you suppose we frame problems in terms of the *proportion* of heads (or toast butter-side down, or whatever) instead of the *number* of heads?
  - ✎ What happens when **p** leaves the range from 0 to 1? Why do you suppose that happens?
  - ✎ What happens to the distribution you get if you add more cases to the collection? What's a good number of cases to have? Why? (1000 is one good number.)

YOU JUST MADE A TOOL. You now have a tool you can use for *any* problem that is functionally the same as flipping coins, even when the probability of heads is not ½. Save this file as **binomialSimulator.ftm** so you can find it again!

9.5 Sampling test scores, I. Here we need a full three-step setup in Fathom. We'll start with the data, draw a sample, then take measures from the *sample.*

❖ Open the file **EX09_05 Test scores.ftm**. Double-click the **test scores** collection to open its inspector. In the **Measures** panel, make a new measure called **scoreBar**. Give it the formula **mean(score)**. Verify that its value is 69.4.

❖ Right-click on the test scores collection and choose **Sample Cases**. The sample collection appears.

❖ Double-click **Sample of test scores** to open *its* inspector. Check out its **Measures** panel: The sample collection inherited the **scoreBar** measure from its source. That value is the mean score of the 10 cases *in the sample.*

❖ Back in the **Sample** panel, change the number of cases sampled from **10** to **4**. Click the **Sample More Cases** button or right-click the collection and choose **Sample More Cases** from the context menu.

❖ Make a case table for that collection so you can see if things are working.

❖ Right-click on the sample collection and choose **Collect Measures**. A new collection, **Measures from Sample of test scores** appears.

❖ Graph the five values of **scoreBar**.

❖ Collect more measures until you have 100 or more. If you do this gradually, and leave animation on, you can see the sampling distribution gradually emerge.

✐ Is the center of the distribution of **scoreBar** close to the mean of **score**?

✐ Does the distribution look more or less Normal?

✐ Does the distribution of **scoreBar** resemble the distribution of **score**?

✐ **ScoreBar** cannot have just any value. Name some impossible values for **scoreBar** between 65 and 75, and explain why they're impossible.

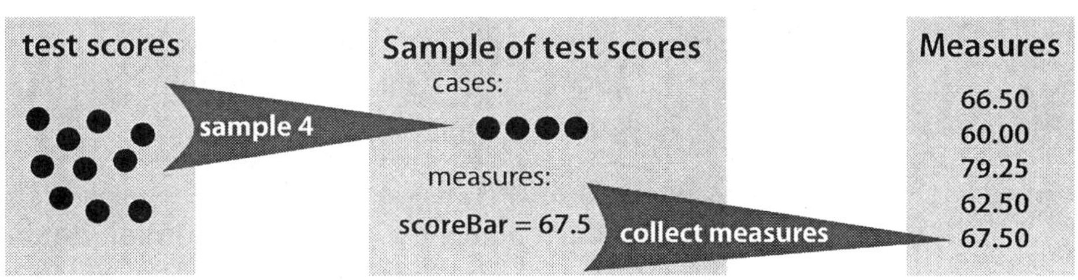

A schematic of our three-collection setup in Exercise 9.5.

Example 9.5 Are you a *Survivor* fan?

Use the **binomialSimulator** you made in Exercise 9.4B (page 130) to get the sampling distribution of $\hat{p}$ quickly.

❖ Set **n** to **100** and **p** to **0.37**. Verify the observations about center and spread of the proportions from the textbook.

❖ Change **n** to **1000**. Verify the changes in the distribution described in the textbook. Remember, to rescale the histogram, just choose **Histogram** again from the pop-up menu in the corner of the graph. To change the bin width, just grab the edge of a bar and drag.

More exercises

9.7 GUINEA PIGS. This exercise works very much the same way as Exercise 9.5 (page 131).

❖ Open the file **EX09_07 Guinea pigs.ftm**. It has the survival times of the guinea pigs. You will not have to label them with numbers because Fathom will do the sampling.

❖ Make a sample of the **guinea pigs** collection. Step (c) says to choose an SRS if size $n = 12$, so change the sample collection to sample 12 instead of 10 (the default).

❖ The sample $\bar{x}$ is a statistic; we will make it a measure. Call it **timeBar** and put it in the original **guinea pigs** collection and then resample.

The value of **timeBar** in **guinea pigs** is really μ, the *population* mean, while **timeBar** in **Sample of guinea pigs** is the *sample* mean $\bar{x}$.

❖ Resample several times and see how **timeBar** changes.

❖ **Collect measures** from the **Sample** collection. Make a graph of the distribution of **timeBar**. Increase the number of measures so you have at least 100. Don't forget that turning off animation makes sampling and collecting measures go faster.

✎ With only 100 measures collected, the graph probably looks skewed to the right. Why is it skewed that direction? Would collecting even more measures make the distribution look more Normal? Try it and see.

9.8 BEARING DOWN. The file **EX09_08 Bearings.ftm** has the data. When you get to step (c), use **Plot Value** to mark the mean. You might want to use **binomialSimulator** from 9.4B to see the sampling distribution for an unbiased distribution of samples of size **n = 200** from a population with **p = 0.1**.

9.14 BAD EGGS. Use **binomialSimulator** from 9.4B again. Consider doing the following:

❖ Rename the collection and the attributes to reflect the context of the problem.

- ❖ Reduce the number of cases in the collection to match the problem as stated. (Think carefully about what corresponds to the number of cases in the collection, and what corresponds to the value of **n**—or are they the same?)

- ❖ Redo the whole thing using the **Collect Measures** mechanism. In step (a), for example, you will have a collection of 10 square yards, and calculate the proportion with egg masses in a *measure*. Then you'll collect those to make your distribution of $\hat{p}$.

9.16 ROLLING THE DICE, II. This is good to do in Fathom.

(a) Make a table with 100 cases. Make four columns: one for each die and one for the **sum**. (Or you could just do one column and make a bigger formula.)

(b) Define a new attribute **dieBar** to be **sum/3**.

(c) Plot a histogram of the **dieBar** values.

9.18 SIMULATING SURVIVOR. You don't need to write the program because you have your **binomialSimulator** from 9.4B! Open that file, set **n** to **5** and **p** to **0.41**.

- ❖ Make a new attribute in the table. Call it **SS** (for sample size) and give it a formula to set it equal to **n**. This records the sample size that each case came from. It looks funny in the table—every row is the same—but we'll use it in a moment.

- ❖ Right-click on the collection and choose **Sample Cases**. A **Sample** collection appears. It has 10 cases, sampled from the original collection.

- ❖ Double-click the **Sample** collection to bring up its inspector. *Very important:* Un-check **Replace existing cases**.

- ❖ Change the sample size **n** to **10**. Right-click on the **Sample** collection and choose **Sample More Cases**. (Ah! Now we have two sets of data in the same collection. Half of it is from **n** = 5, half from **n** = 10, and we can tell which is which because of the values in **SS**.)

- ❖ Graph **pHat** *from the sample collection*. Then drag **SS** to the other axis; be sure that the SHIFT key is down when you drop the attribute. Now you see the two distributions side by side.

  ✐ What happens if you do not hold SHIFT down? Try it and see.

- ❖ Use the same procedure to add samples of 10 with **n** = **25**, **n** = **100**, and **n** = **500**. See how the graph changes.

- ❖ Use a summary table to get the mean and sample standard deviation of each sample. Use the formula **s( )** for sample standard deviation. Again, be sure to hold SHIFT down when you drop **SS** on the "other" axis.

9.2 Sample Proportions

Much of this section is about how, if certain requirements are met, you can approximate the sampling distribution of a proportion with a Normal distribution. We dealt with tools for doing Normal calculations back in Chapter 2, so you can re-use your work from that time.

Because this section is about sample proportions and how a sampling distribution of proportions behaves, you can often use the **binomialSimulator** that you built for Exercise 9.4B to help you. (If you did not do 9.4B, now is a good time.) In particular, you can use it to verify the Normal calculations you do.

Example 9.7 Applying to college

35% of all college students applied to colleges besides the one they are attending. If we have an SRS of 1500 college students, what's the chance that their sample proportion $\hat{p}$ will be within two percentage points of the true value, 35%?

The example in the textbook performs a Normal calculation. How shall we use the **binomialSimulator** to verify the result?

❖ Open the file. Set **n** = **1500** and **p** = **0.35**. If you have 1000 rows in your table, you now have the results of 1000 polls. What proportion are between 33% and 37%? (There are many ways to find this number, though it's easy to forget them. Some ideas: select, point, and look in the status bar; make a measure or a computed slider; use filters.)

Example 9.8 Survey undercoverage?

Again, use **binomialSimulator**. Set **n** and **p** to suitable values. Then see if **pHat** ≤ 0.092 seems unusual. Find the (empirical) probability. Rerandomize to see it change.

❖ If you have trouble telling where 0.092 is on the **pHat** graph, use **Plot Value** to show it.

Easy, huh? But you still have to think. For example, you should be able to answer these:

- ✎ What are the suitable values of **n** and **p**? Why?
- ✎ Why do we ask whether **pHat** ≤ 0.092 is unusual? Why not just "less than?" Why not just "equal to?"
- ✎ What effect does the number of cases in your collection have on the distribution?
- ✎ If you were to do this with measures, how many cases would you have in your source collection? Why? What attribute(s) would the cases have? What measure(s) would you collect?

Exercises

You can use the **binomialSimulator** to check your work with the Normal distribution for practically all of the exercises in this section. An exception:

9.30 RULES OF THUMB. Although you cannot use the Normal approximation, you can make simulations using the measures mechanism.

(a) Make a collection with 1500 cases, 450 of which are Hispanic, and sample 15 for the union executive committee. Or, because 1500 is so large, just make a collection with 15 members for the committee and set it so that the probability of being Hispanic is 0.3. You need a measure for the number (or proportion) of Hispanics; collect it to determine the probability.

(b) Here you need the 316 athletes in a collection; make sure 40% of them have been told to neglect their studies. That's 126.4, so use 126. Then sample 50 for your survey, but be sure to sample *without replacement*.

✎ For both parts, compare the simulated result to the Normal-approximation result. How much difference does it make?

9.3 Sample Means

In this section, the original data—the source collection—is quantitative instead of categorical. When you take a sample, the most common statistic you need is the *mean* instead of the *proportion* of cases that meet some criterion.

That means that the true distribution of samples does not follow the binomial distribution. Nevertheless, as the sample size increases, the distribution of sample means gets closer and closer to Normal, just as the binomial distribution does (given the two rules of thumb). So you will do the same sort of Normal calculations as you did when you studied proportions.

From a simulation point of view, there are similarities as well. You can still use Fathom to simulate the situations in order to check on your Normal calculations and also to be sure you understand what's really going on. If you set up an original collection, take samples, and then collect measures, you don't have to worry about the Normal distribution, or *z*-values, or rules of thumb, or any of that.

Exercises

9.31 BULL MARKET OR BEAR MARKET? Let's see how to use Fathom to illuminate some of the sub-questions. You can use the same techniques in many of the other exercises:

(b) What's the probability that a randomly chosen stock had a return of +5% or better?

❖ Create a collection with a large number of cases and one attribute, **return**, that uses the formula **randomNormal(–3.5, 26)**.

✎ Where did we get those numbers?

❖ Count the stocks with **return** ≥ 5% and figure out the probability.

(c) Do the same for randomly chosen portfolios of 5 stocks.

- ❖ Select the collection and **Sample Cases**. Make it a sample of 5. (It's 10 by default, remember?) Create a measure, perhaps **meanReturn**, to be the average return in the sample.

- ❖ **Collect Measures** from the sample. Increase the number of measures to build up a sizeable (>100) sampling distribution, and find the proportion of samples for which **meanReturn** ≥ 5%.

(d) Use the same distribution to answer this question.

Extension: The data we're using involve the returns of the individual stocks, that is, the percentage they increased or decreased. Is the return of a portfolio of five stocks the same as the mean of the returns of the individual stocks? Consider two things: first, the stocks have different prices; second, the fact that if something increases 50% and then decreases 50%, you have less than when you started.

9.32 ACT SCORES. This is really the same as the previous question but with different numbers. Does the fact that ACT scores are integers make any difference?

The Central Limit Theorem

The Central Limit Theorem (CLT) is one of the deepest wonders of statistics. It lets us treat sample means from any distribution as Normal provided the sample is large enough. Let's spend some time playing with sampling distributions. We will find out how Normal the sampling distributions are for various source distributions and sample sizes, and whether the resulting mean and spread are about the same as what the CLT predicts.

- ❖ Open the file **CLT Playground.ftm**. It's in the folder for Chapter 9, and should look like the illustration.

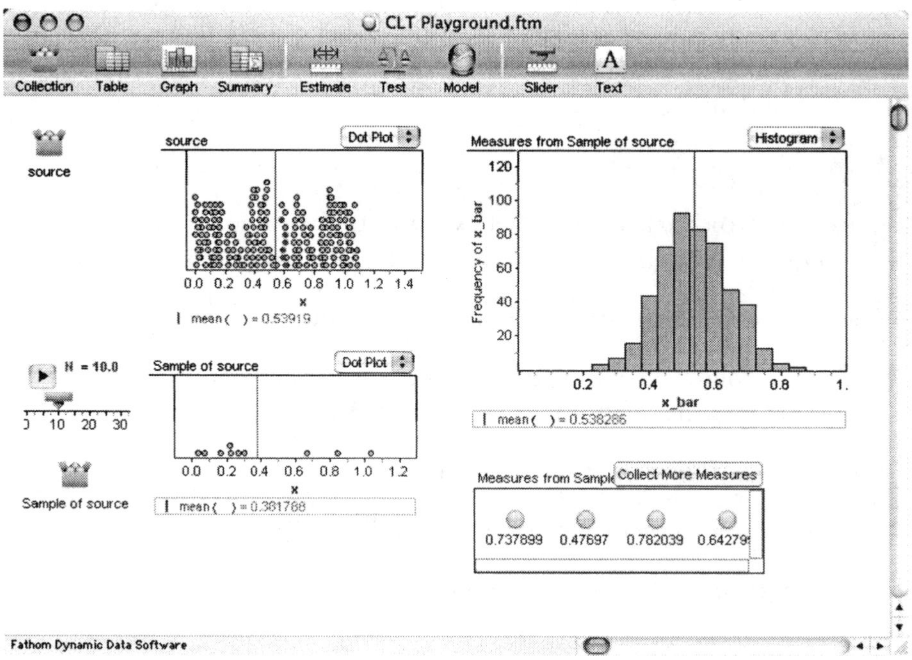

You'll see the **source** collection in the upper left. It has 200 cases, each with a value for **x**. Below it is the **Sample of source** collection; you control the sample size with the **n** slider. On the right is **Measures from Sample of source**. It collects **x_bar**, the mean of **x** in each sample. When you press the **Collect More Measures** button, Fathom resamples the **source** collection 500 times, collecting **x_bar** and plotting the results in the histogram. (If it's too slow, reduce the number of measures you collect.)

The distribution looks pretty Normal. Your challenge is to change things so that the distribution of **x_bar** is obviously *not* Normal. Do so in as many ways as you can so you get good at telling where the CLT does not apply.

You can only control two things right now: the sample size and the original distribution.

- ❖ Change that sample size by changing **n**. What if you set **n** = 1? Then what does the distribution look like?
- ❖ Change the distribution by selecting and dragging cases in the upper-left dot plot. You can drag boxes around whole groups to select them. One suggestion: Drag them into two piles. How does that change the sampling distribution? What if the piles are unequal?

> NOTE. We have used some fancy techniques in making this file. For example, **n** is an *integer* slider, and the three graphs have their axes *linked*.

Enhancing your CLT Playground

Suppose the sampling distribution looks Normal. Is it as the CLT predicts? Let's plot the predicted Normal curve on the sampling distribution. We want it to look like this:

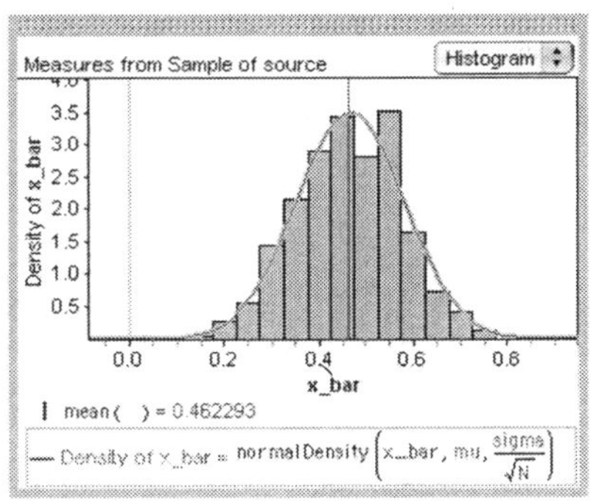

- ❖ First, we need the parameters for our Normal curve. We'll use computed sliders. Drag a slider off the shelf and name it **mu**.
- ❖ Drag the *name* of the source collection—the word **source**, not the box of balls—onto the **mu** slider. Now the slider "knows" about data inside **source**.
- ❖ Double-click the slider axis to inspect it. Enter **mean(x)** into the formula box for **value**. The slider should show the mean of **x**. Since **source** is our population, the **mu** slider really is μ, the population mean.

❖ Use the same technique to make a **sigma** slider, whose formula is **StdDev(x)**. If you can't get it to work, you probably forgot to drop the name of **source** onto the slider.

❖ We're going to plot a Normal probability density curve, so we have to change the histogram to be a density plot instead of one of frequency. Right-click on the histogram and choose **Scale>Density**.

❖ Right-click again and choose **Plot Function**. Enter the relevant Normal function:

normalDensity(x_bar, mu, sigma/sqrt(n))

The function should appear on the graph.

You could make additional graphs (such as a Normal quantile plot) and perform additional calculations (such as comparing the sampling distribution's standard deviation to the predicted one) as you explore the CLT Playground.

Example 9.13 Servicing air conditioners

Sometimes we don't have to set up the whole source-sample-measures mechanism. In this problem, we can use a random-number function to take the place of sampling.

A good function to use is **randomExponential(1, 1)**. We have set that up in the file **EG09_13 Air Conditioners.ftm**, in a collection (**service times**) with 70 cases. That represents the 70 air conditioners. That collection itself serves as a sample of all possible air conditioners, so you do not have to sample.

You need to create the measure and collect it appropriately in order to solve the problem.

❖ Check the mean and standard deviation of the **service times** collection. Are they both close to 1.0?

Exercises

For these exercises, you mostly need to perform Normal calculations. Use the skills you developed back in Chapter 2—but be sure you can do these without a computer. It is interesting to simulate any of these using sampling or random-number generation; compare your results to the Normal calculations. Here are a few notes and additional ideas:

9.35 BAD CARPET. The problem states that the population distribution cannot be Normal because a count has only integer values. True!

❖ Create a collection of data with integer-only values that has a mean of 1.6 and a standard deviation of 1.2. Come as close as you can with 10 cases.

❖ Sample from that collection, and collect measures. See if the answer to the question you get from simulation corresponds to your CLT answer.

9.36 INVESTING FOR RETIREMENT. Simulate this in Fathom.

- ❖ Make a collection with 40 cases (one for each year) and one variable, named **return**, to represent the return that year.
- ❖ The problem states that the mean annual return is 13.2% and that its standard deviation is 17.5%. That tells you what formula to use.
- ❖ You will need to collect a measure. (You have to figure out what that measure is.) Collect at least 1000 of them so you can estimate the probabilities the problem asks for.
- ❖ Extension: Figure out what Andrew's *total* return is likely to be.
  - ✐ What is the chance that his total return is less than 10%? Less than 100%? Less than 1000%? Less than 10,000%?
  - ✐ You cannot figure out the total return from the mean annual return. Why not?

9.43 WHAT A WRECK! As you did in Exercise 9.35, construct a distribution of *integers*. This time you need a mean of 2.2 and a standard deviation of 1.4.

9.48 BAGGAGE CHECK! We're back in "proportion land" now, so **randomBinomial( )** is a useful function. You could use **randomBinomial(100, 0.7)/100** to get the proportions for part (a). Use 50 cases.

9.49 IQ TESTS. You can simulate this in Fathom and compare your results to what you get using your understanding of Normal distributions.

Note that to answer (b) and (c), you need a sampling distribution. That means you'll need to collect a measure.

9.51 DETECTING GYPSY MOTHS. See if you can make a distribution that has the properties described in the textbook. What is its median? What is its *mode*?

- ❖ Sample from the distribution to answer (a) and (b) using simulation.

Some final words on simulation

Ever since Chapter 6, we have been studying probability, randomness, and distributions. In the textbook, this has culminated in the study of sampling distributions and how, under common conditions, these are approximately Normal with means and standard deviations we can calculate easily. In the next few chapters, we will start to use all of this reasoning backwards and infer things about a population based on a sample. The text uses the Normal machinery, calculating probabilities using Normal functions, especially cumulative density and its inverse.

As you might have gathered, you can get the same results through simulation. If you simulate properly, by taking samples, devising measures, collecting them, and then looking at the resulting sampling distribution, you do not need to worry about any of the limitations, caveats, and rules of thumb that make the Normal approximation valid. For example, suppose instead of the mean of a population you're interested in the maximum value. That statistic—maximum—does not follow the Central Limit Theorem, so its sampling distribution is not Normal (try it!) and you can't use a Normal approximation. But you can use our "measures" machinery just fine.

Why don't we just use simulation for statistics? The historical reason is that the computing power you need in order to simulate effectively has only recently been available. Another reason is that the whole Normal-distribution thing is quite lovely and general and lets you see connections you might otherwise miss. Finally, if you're on a desert island and you have only your graphing calculator, it may be hard to set up a simulation and run off a few thousand cases. **Normalcdf**, or even just the 68–95–99.7 rule, may be all the horsepower you need.

What do they do out there in professional statistics? Nowadays, they do both: They simulate *and* they use traditional techniques. So learn both: They will reinforce one another.

10 Estimating with Confidence

10.1 Confidence Intervals: The Basics

Much of the beginning of this chapter is vital conceptual stuff that does not need Fathom. Two exercises, however, review important old material about combining random variables and least-squares lines. We'll use these exercises to do that review and to introduce important Fathom techniques.

10.6 TALLER GIRLS. We're supposed to find the probability that the mean height of 10 girls is greater than the mean height of 7 boys. The girls are from a Normal distribution $N(54.5, 2.7)$ and the boys are from $N(54.1, 2.1)$.

Unlike previous exercises in this chapter, part (b) of this exercise asks you to simulate the data to estimate a probability. Although you have learned many simulation strategies in Fathom, this problem has a pitfall that's worth learning about.

You might try to make two collections—one for the boys and one for the girls. That's great until it comes time to compute the probability. You will need to collect measures somehow, but there is no way (yet) to collect measures from two collections simultaneously. We therefore need a *single* collection with all 17 students.

❖ Make a new table and give it two attributes, **sex** and **height**.

❖ In the **sex** attribute, make 10 girls (**F**) and 7 boys (**M**).

❖ Make a formula for the **height** attribute, one that uses **sex** to determine **height**. Be careful to enclose the **F** and **M** in quotes, as in the illustration:

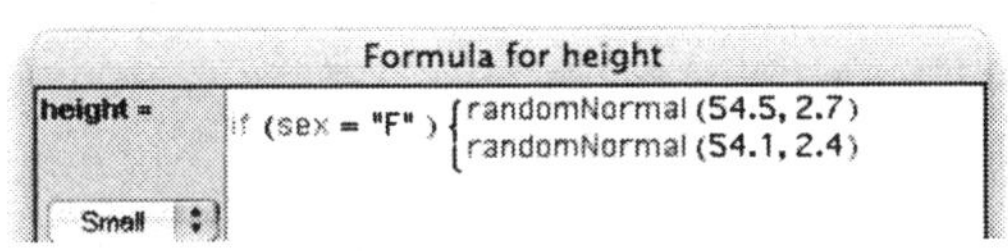

❖ Make graphs and rerandomize to make sure things are working right.

❖ Create a measure, **tallerGirls**, with a formula such as **mean(height, sex = "F") > mean(height, sex = "M")**. This is a Boolean expression, either **true** or **false**.

IMPLIED "FILTERS" IN FATHOM. See how we put mini-filters in the formulas with **mean**? They go inside the parentheses and after a comma. You can use them with any function that works on a set of cases. So, for example, you could write **stdDev(height, age > 80)** to study the spread of heights of old people. But writing **cos(angle, base > 5)** doesn't make sense.

❖ Collect at least 100 measures and look at the distribution you get. What does it tell you about the probability you're looking for? Your document will look something like the one in the illustration.

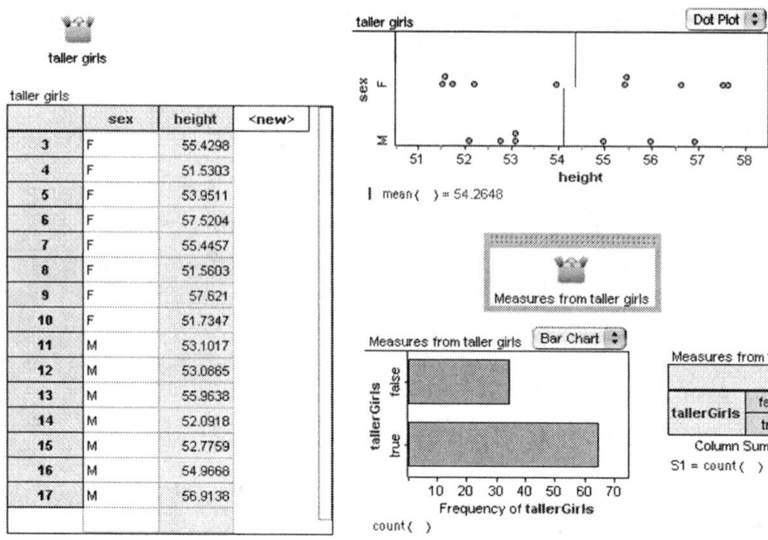

ESTIMATION AND ESTIMATION. We using the word *estimation* in two senses. In this chapter, an *estimate* is our best guess as to a population parameter, and we'll see that this estimate will be an *interval.* For example, our estimate of the mean height of *all* girls is the mean height of this SRS of 10 girls (plus or minus something). But in probability problems, we do simulations to *estimate* the answer; in this case we mean that we're not doing a theoretical calculation—we're looking at the situation empirically, and if we were to do it again, we'd get a different answer.

10.12 SISTERS AND BROTHERS, EXTENSION. After you have done the problem, you have a **sister-brother** graph with a least-squares line (LSRL).

What's the best way to do part (c)? One strategy is to plug 70 into the LSRL equation. But Fathom offers a graphical alternative.

❖ Point at the line; a red dot appears. Slide it to the right place and read the number in the status bar. See the illustration.

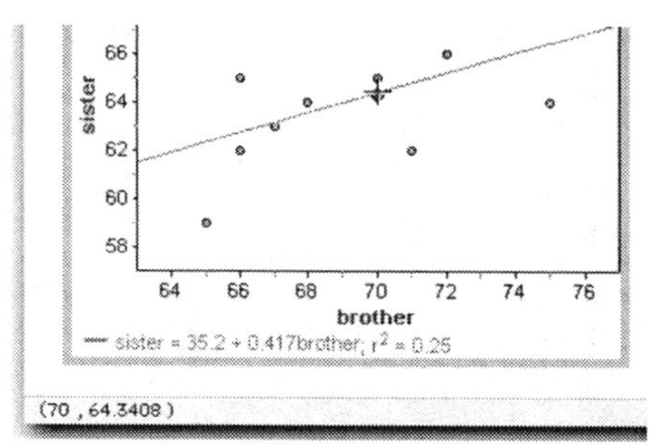

Now explore the variation in the linear model and in the prediction you make for Tonya's height:

- ❖ Make a new collection with two attributes, **slope** and **prediction**. In the first case, enter the slope of your LSRL and the prediction you made using the data. These will be something like 0.417 and 64.3.

We're going to use a technique called the *jackknife*. The idea is that the points we have are just a sample. What would the line be like if we had a different sample? We can't actually get more data, so to see what a different sample could look like, we'll see how the graph changes when each point is *missing* from the data set.

- ❖ Select one case and **Cut** it (CONTROL-**X** or ⌘-**X**). The point will vanish and the line will change. Record the slope and your new prediction.
- ❖ **Paste** (CONTROL-**V** or ⌘-**V**) the case back into the graph.
- ❖ Repeat the process, recording **slope** and **prediction** with each point missing.

Now you can graph **slope** and **prediction** (separately) and see what those distributions tell you. The **slope** graph anticipates Chapter 15; this is a legitimate way to tell how well you know the slope of a least-squares line.

The **prediction** graph, similarly, tells you how much your prediction would vary if you had slightly different data to help you make that prediction.

But notice that the spread of your **prediction**s is probably quite a bit less than any spread you might have predicted in part (d) of the exercise in the textbook. To answer part (d), it's reasonable to look at the spread of the points around the line. The range of plausible heights for Tonya extends from about 61 to 68 inches (see the illustration below). But the spread in **prediction**s on our graph extends only from 63.8 to 65.

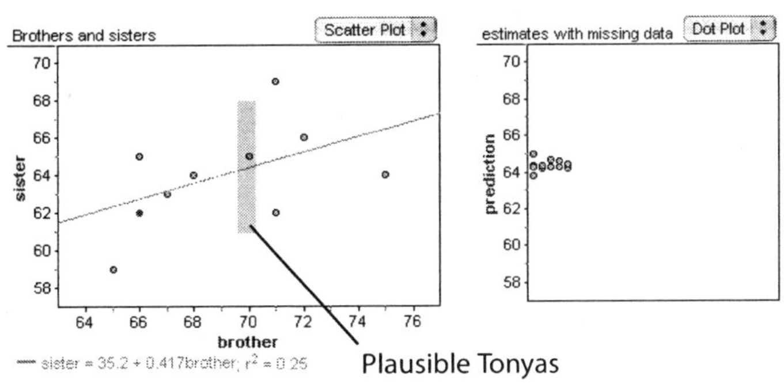

Why is the spread so much smaller? The reason is deeply connected to the whole point of this chapter.

In a universe of Damiens and Tonyas, the heights of possible Tonyas themselves vary quite a bit—plus or minus three inches at least. That's the spread of points in the graph about the LSRL. Your **predictions**, on the other hand, are estimates of something equivalent to a *parameter*: the *mean* of the Tonyas. And how well do we know that? That's more like plus or minus one.

IT'S IMPORTANT TO GET THIS STRAIGHT. The range you get with a confidence interval is not the spread of the data. Rather, it's the spread of a sampling distribution of some statistic (often the mean). The range is the uncertainty in that statistic, not the variability in the data.

More confidence interval basics

As you have read in the text, getting a feel for how confidence intervals behave is really important. To get that "feel," you experiment with a confidence-interval applet and do examples that show you how sample size and confidence level affect the size of confidence intervals.

We will do some of that now in Fathom, beginning with how confidence intervals behave as you change other parameters. This section assumes that you have already done the appropriate material in the textbook (section 10.1) and are using this to reinforce your study.

The key issues are sample size and the spread of the data within the population. We will assume here that the population mean μ is 10.0, though you can change that using the **mu** slider in our document.

❖ Open the document **confidence intervals I.ftm**. There you'll see simulated data, a confidence interval (represented as the region between a pair of lines on a plot), and sliders that control many aspects of this simulation.

❖ Move the sliders and see what happens. Investigate the following:

- ✐ How does changing **sigma** change the sample? How does it change the confidence interval?
- ✐ How does changing **n** change the sample? How does it change the confidence interval?
- ✐ How does changing **C_level** (the confidence level) affect the sample and the confidence interval?
- ✐ Suppose you wanted a confidence interval with a margin of error of about 0.1. If you have a 95% interval (**C_level** = 0.950), how big does **n** have to be? Move the sliders to figure it out, then check your result with calculation. (Leave **sigma** at 1.0.)
- ✐ Suppose you had a sample size of 100. What confidence level would you have to accept to get a confidence interval with a margin of error of about 0.1? Again, do it with sliders, then check.

Here are additional, challenging things to consider:

- The formula for **zStar** in its computed slider is

  zStar = normalQuantile((1 + C_level)/2).

  Why isn't it just **normalQuantile(C_level)**? That is, what does adding one and dividing by two accomplish?

- Why do we use the **normalQuantile** function instead of **normalCumulative** (besides that one works and the other doesn't)? That is, the two functions are closely related. What is the underlying difference between them?

Exploring that file shows what happens with a single confidence interval. But the tricky part of the confidence concept is that if you construct *repeated* confidence intervals, some of them don't capture the true mean. Let's take many samples and see:

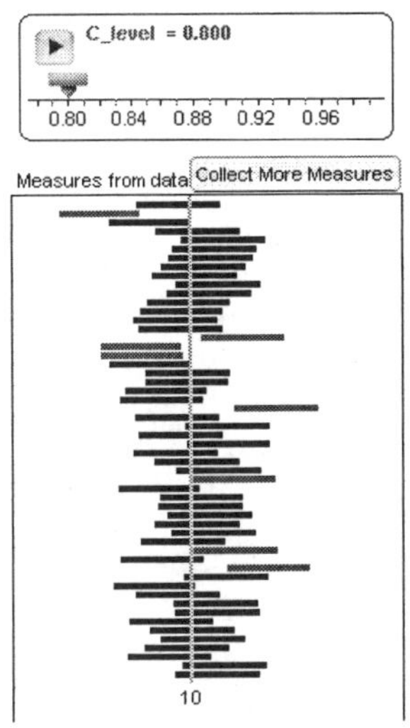

- ❖ Open the file **confidence intervals II.ftm**. This file shows a set of confidence intervals from different samples and lets you see how these intervals change as you change **C_level**.
- ❖ Press **Collect More Measures** to get five additional confidence intervals. They are color-coded by whether the interval captures the true mean or not.
- ❖ Change the **C_level** slider to see how the intervals change, and how more or fewer intervals get captured.

REMEMBER, YOU DON'T KNOW THE MEAN. We have set the true, population mean to 10. That way we can take a sample and see whether its confidence interval captures that true mean. But in real life, you *never know the true mean*. So as you do these activities, imagine what it would be like to be trying to figure that mean out—with only *one* sample. Later, when you really have only one sample, do the opposite: imagine what it would be like to get many samples. Does your confidence interval encompass the true mean? Maybe, maybe not.

10.2 Estimating a Population Mean

Remember how the textbook kept saying, "In real life we don't know the population standard deviation?" Fathom only calculates confidence intervals in the realistic setting—and that's what we're going to learn about in this section.

As the textbook explains, we use s, the sample standard deviation (which we have) instead of σ, the population standard deviation (which we don't have). When we divide it by $\sqrt{n}$, we find that its distribution is not quite Normal; it has what's called a t distribution. In Fathom, the t functions are under **Distributions** in the formula browser, under **Student's t**.

Activity 10C Comparing the *z* and *t* distributions

We will go through this activity in Fathom, building the display from scratch. We'll use the step numbers from the textbook so you can see what corresponds to what.

1. Start with a blank Fathom document. Make a new empty graph and choose **Function Plot** from the menu in the corner.

2. Choose **Plot Function** from the context menu, and enter **normalDensity(x)** in the formula editor. You'll see a thick blue line with a little bump.

4. (We'll do this step before step 3 so you can see the plot…) Rescale the axes so that you can see the Normal shape.

3. Choose **Plot Function** again; enter **tDensity(x, 2)**. Now you'll see both curves. If you click on one of the functions at the bottom of the graph, its graph gets thick.

5. (and 6) The **2** in the **tDensity** function is the number of *degrees of freedom*. How does changing the degrees of freedom affect the graph?

   - Make a new slider and rename it **df**.
   - Edit the second function to be **tDensity(x, df)**.
   - Change **df** to see how the function changes.

THE MAGNIFIER. To study the distributions more closely, it helps to zoom in. Hold down CONTROL (Windows) or OPTION (Mac) and point at the graph. The magnifier cursor appears. Drag a rectangle around the "foothills" of the graph, as shown. When you release, the graph expands and you can see the details more clearly.

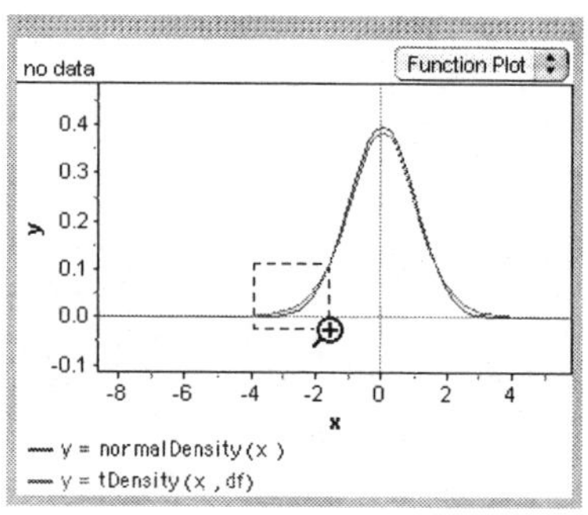

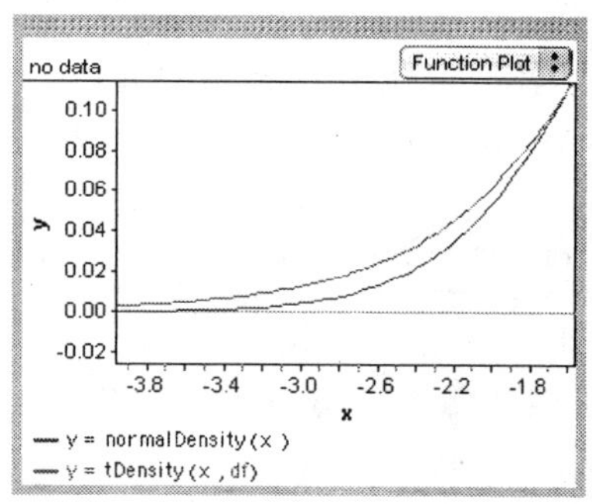

Example 10.9 Auto pollution

Let's use this example from the textbook to learn to use Fathom's confidence-interval tool.

- Open the file **EG10_09 Auto Pollution.ftm**. It has the data from the text.
- Let's assess Normality. Make two graphs; in one, put **NOX** on the horizontal axis and make it a box plot; in the other, put **NOX** on the vertical axis and make it a Normal quantile plot. Your graphs should look like the ones in the illustration below.

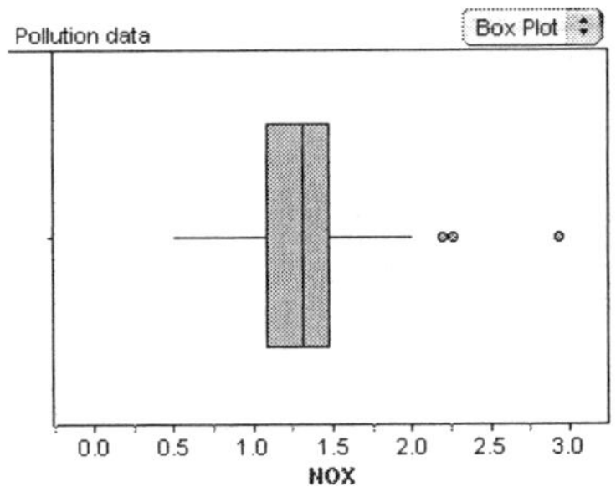

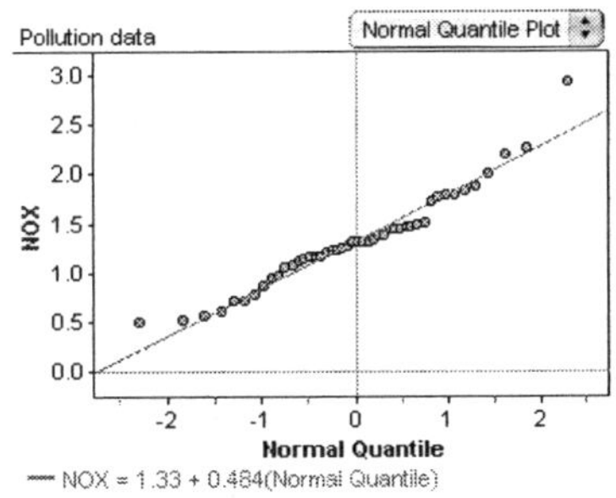

Note the outliers as described in the textbook. Also, in the Normal quantile plot, look at the formula for the line. When variable is on the *vertical* axis, the intercept is the mean and the slope is the standard deviation. (You can verify this with a summary table.)

- Now for the calculations. Drag a new *estimate* from the shelf. Estimates, like graphs, start out empty. You can change what they do using a menu, and you can drag variables to them.

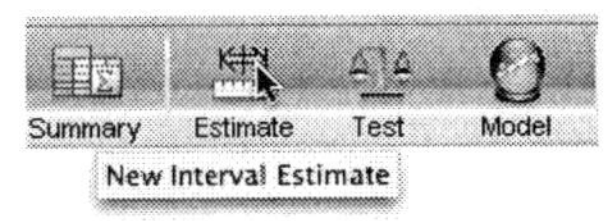

- Use the menu in the corner of the estimate (it's just like the menu in a graph) to change it from **Empty Estimate** to **Estimate Mean**.
- Drag **NOX** to the top area where it says **Attribute (numeric): unassigned**. The estimate fills in with information from the collection, and tells you, in detail, about the confidence interval. If you don't like that much text, un-check **Verbose** in the context menu. The regular and terse versions of the estimate appear in the illustration:

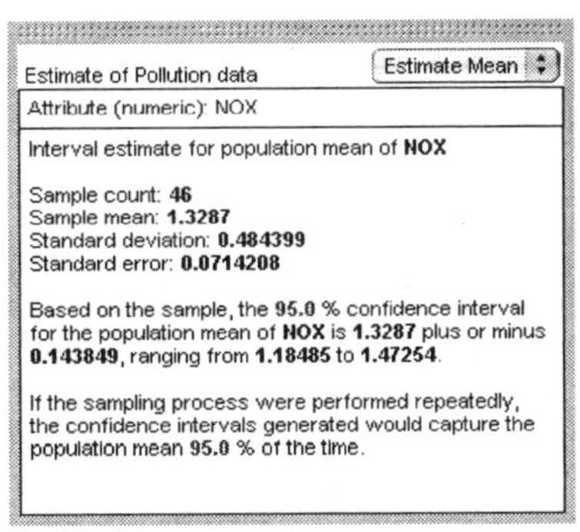

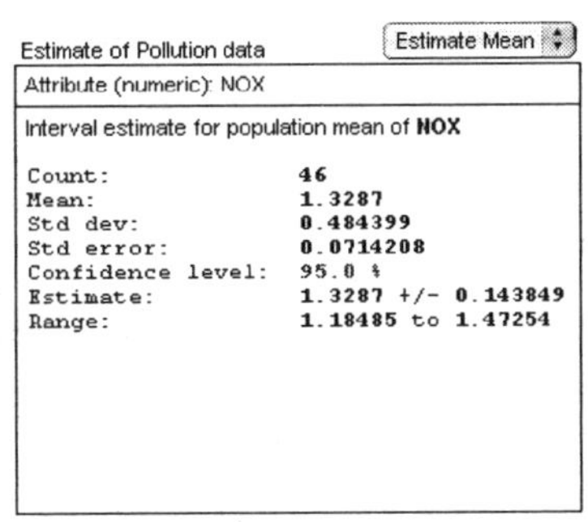

The paragraph at the bottom of the regular estimate has in it the True Meaning of Confidence Intervals, which gives people a lot of trouble. Study it.

Two more things to know right away about estimates in Fathom

First, make another estimate of a mean, but don't drag **NOX** in. Notice that there are blue numbers all over the place. These are *editable*. Why?

Suppose you know the sample size and the mean and the standard deviation of the sample, but you don't have all the data. You can still have Fathom compute a confidence interval. Just edit the values and the interval appears. *This is how you can do many exercises.*

For example,

10.27 Standard error. We'll start with part (a):

❖ Make an "Estimate of Mean": drag an estimate off the shelf and use the corner pop-up menu to change it to **Estimate Mean**.

❖ Edit some blue text: change **attribute name** to **systolic**, **Sample count** to **27**, **Sample mean** to **114.9**, and **Standard deviation** to **9.3**. As you edit, the results change; you can just read the standard error in the text.

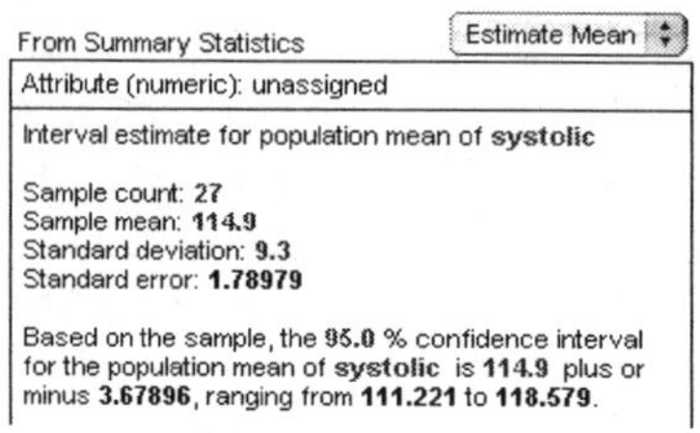

In part (b) you get standard error and have to calculate standard deviation; this Fathom technique won't help with that. You have to resort to the formula $SE = s/\sqrt{n}$.

Second, if you right-click on one of the blue numbers, you can choose **Edit Formula**. If you use slider values in your formulas, you can see how varying the confidence level, sample size, standard deviation, etc., affects the confidence interval.

Hard to get the right-click to work? It may help to left-click in the number or near it first, to get the cursor there, and then right-click.

Example 10.10 Is caffeine dependence real?

Let's see how to do this paired-*t* example in Fathom. We'll focus on the change in "beats" scores. You can use the same techniques to check your work on the depression.

❖ Open the file **EG10_10 Caffeine study.ftm**. It contains the data from the textbook.

❖ Make a new column, **Beats_diff**. Give it a formula, **Beats_caff – Beats_plac**. (We could do it the other way around, but this way the difference is the *increase* in beats per minute under the influence of caffeine.)

❖ Graph the difference to assess Normality. (Use a box plot and Normal quantile plot.) You will see that there are serious outliers, so our subsequent results may be suspect; you'll have to decide what's right in Exercise 10.34.

❖ Pressing on anyway, drag a new estimate off the shelf, change it to **Estimate Mean**, and drag **Beats_diff** to the box that reads **Attribute (numeric): unassigned**.

❖ Edit the **95.0%** confidence level to read 90%. You should find a confidence interval ranging from about –6.5 to about +47.

❖ To see how the two upper outliers affect your results, select them (draw a box around them in the box plot) and **Cut**. See how the analysis changes? For one thing, the mean difference gets very close to zero. **Paste** the data back in before you close the file.

The key step in the paired procedure is to create a new column and calculate the difference between the two variables you're comparing. Then you do a *one*-variable analysis on that difference.

You might have noticed that the estimate object has other options besides **Estimate Mean**; you could also choose **Difference of Means**. This is for making an *unpaired* estimate. For example, if you had the weights of a sample of ostriches before and after a new diet, *but you didn't keep track of which ostrich was which*, you would use this to estimate the difference in the mean weights. If you use this for the Beats data (you'll drag **Beats_caff** and **Beats_plac** to the *two* slots in the box), you will see that the mean difference is the same (it has to be) but that the interval is wider.

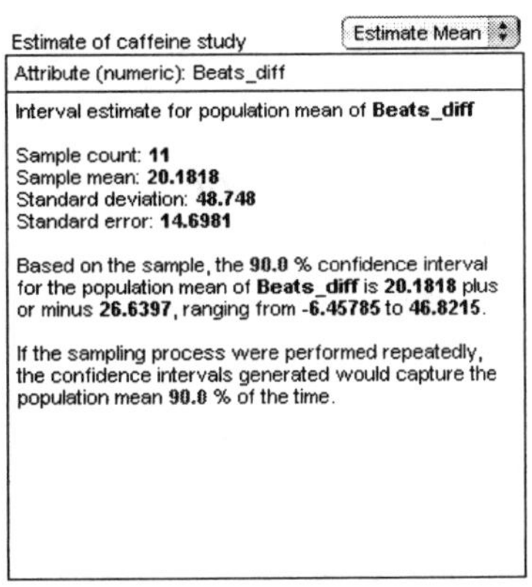

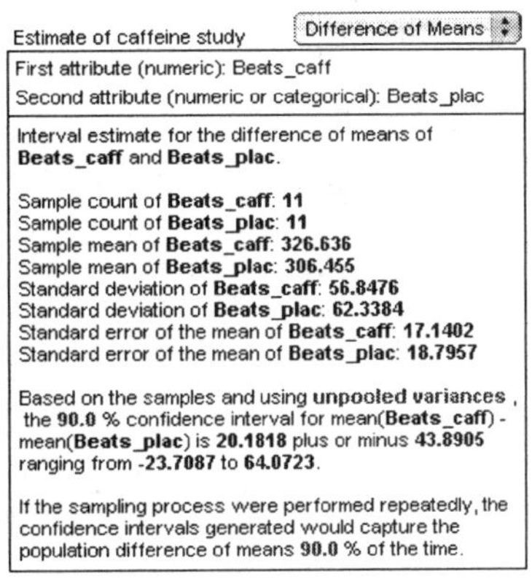

Exercises

Here is a series of straightforward exercises for which we have the data:

10.33 Does playing the piano make you smarter? Use **EX10_33 Piano smarter.ftm**.

10.35 The moon made me do it! Use **EX10_35 Moon made me do it.ftm**.

10.38 Red wine is good for the heart. Use **EX10_38 Red wine.ftm**.

10.39 A big-toe problem. Use **EX10_39 Big toe problem.ftm**. This same file also works for Exercise 10.40.

10.41 Spanish teachers' workshop. Use **EX10_41 Spanish workshop.ftm**.

10.44 Does nature heal best? Use **EX10_44 Newt healing.ftm**.

10.3 Estimating a Population Proportion

In the estimate, you could also choose **Estimate Proportion** or **Difference of Proportions**; these are the subject of this next section. The textbook begins with an overview of sampling variation, to remind you how repeated polling will give you different proportions, but that these proportions follow a binomial distribution—approximately Normal as n gets large—with properties we know. Much of the logic and many of the calculations for estimating proportions are identical to the corresponding logic and calculations for estimating means.

The Fathom techniques are similar as well. In the estimate object, choose **Estimate Proportion** and either drag an attribute to the right place or type in the numbers.

Example 10.13 Binge drinking in college (continued)

We'll start with this example. We don't have case-by-case data, so we'll just type in the results.

- ❖ In a new Fathom document, make a new estimate and choose **Estimate Proportion** from the pop-up in the corner.
- ❖ For **Category**, enter **Binge drinkers**; for **Attribute**, use **Students.** (Using good descriptive labels will make the text make sense and help you understand the results.) Then enter the numbers and the confidence level, which is 99%. Your estimate should look like the one in the illustration:

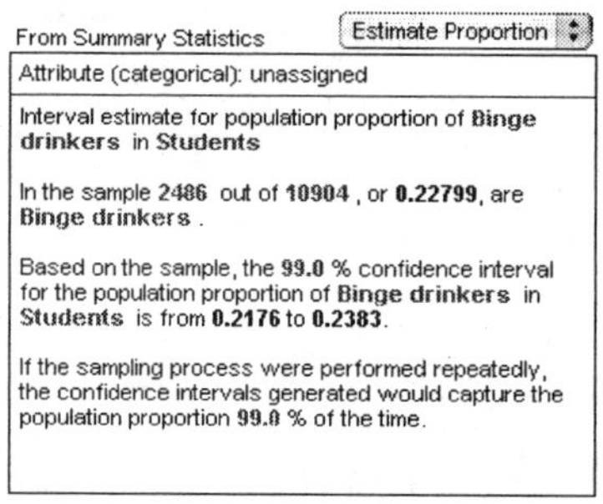

You can use this tool to help you with many exercises. And as before, you can use sliders to control values if you want: just right-click on the blue number and choose **Edit Formula.** We'll use this technique to do the next example.

Example 10.17 Customer satisfaction

We want to see how big a sample size we need to estimate p to within 3% at a 95% confidence level. *You need to know how to do this without a computer*, but doing this with a computer may illuminate what's going on:

- ❖ Make a new estimate and choose **Estimate Proportion.**
- ❖ Make a new slider and call it **n**.
- ❖ In the estimate, enter **Satisfied** for **Category** and **Customers** for **Attribute.**

Now we need to enter the numbers. The confidence level is 95% by default, so we'll leave it alone.

❖ Right-click on the **20** (It's the sample size. How do we know that?) and choose **Edit Formula**.

❖ In the editor (called **formula for sampleCount**), enter **round(n)** and close it. Using the **N** slider, increase the sample size from 5 to something more reasonable, such as 100.

What do we do about the 10? The textbook suggests using $\hat{p}$ = 0.5, so…

❖ Right-click on the **10** and choose **Edit Formula**. The editor is called **formula for countInCategory**; enter **round(n/2)** and close it.

WHY ROUND? The slider can easily give you a non-integer, and Fathom hates the idea of a sample size not being a whole number. Another alternative is to force the slider to be an integer slider, which you can do in its inspector.

❖ Increase the sample size until the interval extends from 47% to 53%, because that's ±3%. Compare your result to what you get when you use the by-hand technique.

Extra: When would you drag an attribute in?

You have been typing in the numbers, but you don't have to if you have the case-by-case data. Let's do an example:

❖ In a new Fathom document, choose **File>Import>U S Census Data**. When the inspector appears, click **Download Data** to get a (more or less) SRS of 500 U.S citizens.

❖ Make a table so you can see some data.

❖ Drag an estimate off the shelf and choose **Estimate Proportion**.

❖ Drag **Sex** from the table to the estimate where it says **Attribute (categorical): unassigned**. The estimate fills with calculations.

Chances are there are not exactly 250 males and 250 females. Does that mean that one sex is dominant in the population? Probably not. In the illustration, you can see that the confidence interval for the proportion of males overlaps 50%.

people

	Census...	State_Fl...	Age	Sex	R
1	2000	Florida	41	Female	Bl
2	2000	Florida	14	Female	W
3	2000	Florida	47	Female	O
4	2000	Florida	2	Female	W
5	2000	Florida	80	Male	Bl
6	2000	Florida	26	Female	Bl
7	2000	Florida	41	Female	W

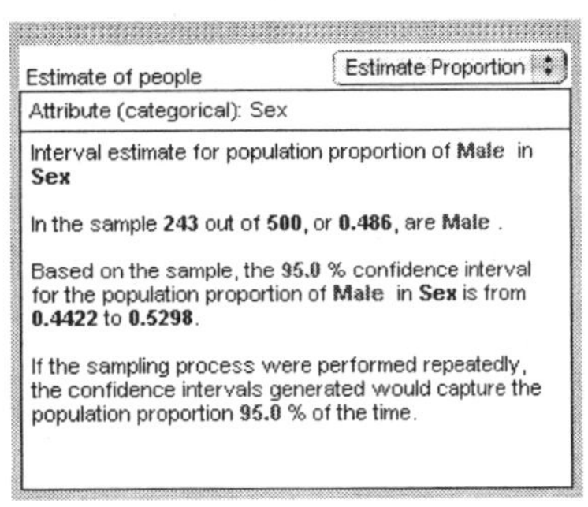

Explore other attributes. If you want to change which category you're looking at (suppose you wanted the proportion of **Females**; or **Divorced** people if you're looking at **Marital_status**, for example), click on the category in the estimate itself. A menu pops up with the different possibilities.

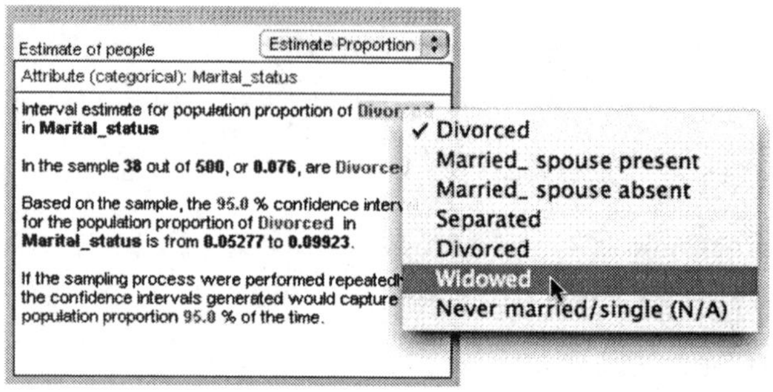

Exercises

10.57 RUNNING RED LIGHTS. Use Fathom to construct your confidence interval.

- ❖ In a new document, drag an estimate off the shelf. Choose **Estimate Proportion** from its menu.
- ❖ Edit the blue text: change **Category** to **Runners** and change **AttributeName** to **Responses**.
- ❖ Then change the next statement to read, "In the sample, **171** out of **880** are Runners." Fathom fills in the rest of the information; you can read your confidence interval from the text.
- ❖ If you were asked for a different confidence level (i.e., not 95%) you could have changed that number as well.

10.60 CAN YOU TASTE PTC? Use a Fathom estimate and a "guess and check" strategy to solve this problem.

- ❖ Make a new estimate, choose **Estimate Proportion**, and enter reasonable terms for **Category** and **AttributeName**.
- ❖ For the numbers, start out with something sensible such as 75 out of 100.
- ❖ That will give a confidence interval that it too wide, so change the numbers—always keeping the proportion at 75%, until you get a ±4% interval.
- ❖ If you're really sneaky, you can use a slider for the sample size, and use two formulas for the numbers.

Special topic: Bootstrap intervals

Computing the confidence interval for the mean has a number of prerequisites, some of which have to do with Normality, or at least having enough points in your sample that the CLT will assure you that the distribution of sample means will be roughly Normal. All of the reasoning hinges on Normality, although, to be fair, confidence intervals work pretty well even when the criteria are not quite met.

Still, real data is rarely actually Normal, and the reasoning is making the assumption, based on limited data, that the underlying population is Normal with such and such a mean and standard deviation.

There is an alternative called the *bootstrap*. It's impractical without technology,

The idea is strange and simple: the only information we have about the population is our sample. So let's assume that the population distribution is *identical* to that of the sample.

❖ Open a file we've used before: **EG10_09 Auto Pollution.ftm**. In it, we found that the 95% confidence interval for **NOX** was 1.185 to 1.472.

❖ Right-click the **Pollution data** collection and choose **Sample Cases**. A new collection appears, called **Sample of Pollution data**. This sample has 10 cases by default.

❖ Double-click the new collection to open its inspector. It will open to the **Sample** panel. Set it to sample 46 cases with replacement. Turn animation off. Press **Sample More Cases** to get a new sample.

We use 46 because that's how many are in the original collection. When we sample, we're drawing randomly from a population that's identical to the original data. But we need to have the same sample size, since that affects how much the sample mean varies. So we sample with replacement until we have the same number of cases. When we're done, we'll probably have duplicated some cases and left some out. As a result, the mean will vary.

❖ Close the inspector and drag the edge of the sample collection to open it up a bit. You should see a **Sample More Cases** button.

❖ Graph **NOX** from the original collection and **NOX** from the sample, and use **Plot Value** to show where the means are. Press **Sample More Cases** to see how that mean varies.

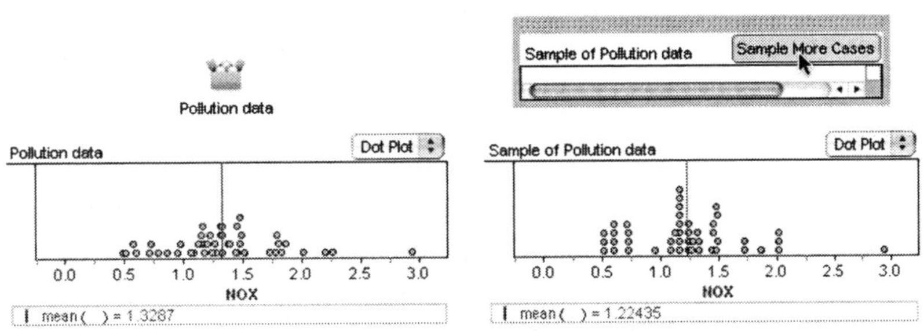

Now we need to collect a lot of those means; this is a job for measures. You can probably anticipate the source-sample-measures structure; the difference from before is the way we set up the sampling so the sample was the same size as the source.

- Define a measure, **NOXbar**, in the sample collection. It's **mean(NOX)**.
- Right-click the sample collection and **Collect Measures**. Verify that the resulting collection, **Measures from Sample of Pollution data,** has **NOXbar**s in it, and that they aren't all the same.
- Collect at least 100 measures. For this purpose, 400 is a great total number. Graph **NOXbar**.

You can see how much **NOXbar** varies; you would be surprised if the true population mean were far outside that distribution. Let's find out where the middle 95% of the data are. To do that, we want to find the 2.5 percentile and the 97.5 percentile of the data.

- Use **Plot Value** to put the two lines on the graph; use Fathom's **percentile( )** function. For example, the first one is at **percentile(2.5, NOXbar)**. You'll see something like this:

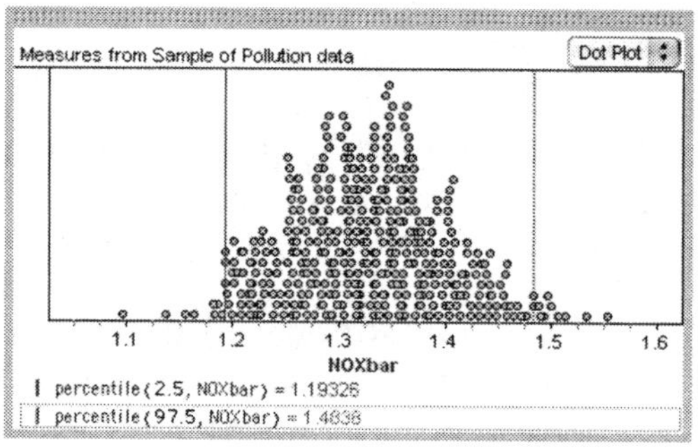

Since 2.5% of 400 is 10, the lines show up at the tenth case from either end. The middle 380 cases—95%—are inside the interval.

Our 95% "bootstrap interval" extends from 1.19 to 1.48, almost identical to the "orthodox" confidence interval (1.185, 1.472). One beauty of this technique is that it makes no assumptions whatsoever about Normality, only that the population has the same distribution as the sample, whatever its shape. A disadvantage of the technique is that you have to create your sampling distribution every time from your data.

The bootstrap is an example of a *nonparametric* procedure. In more orthodox statistics, when we reason based on assuming that data are distributed Normally, we use means and standard deviations. These are *parameters* of distributions. All of the inferential procedures in this chapter and the ones that follow rely on those parameters; for that reason, they're called *parametric* procedures. They are useful and elegant. Until computers, they were absolutely essential. But now that you *can* produce a new sampling distribution every time you need one, parametric procedures are not the whole story.

11 Testing a Claim

The logic of hypothesis testing is similar to that of confidence intervals. The mechanics in Fathom are similar, too: instead of dragging an estimate off the shelf, you'll use a test, but you still drag the relevant variables to the slots at the top and/or edit the blue numbers just as you would with estimates.

This means you already know most of what you need for many of the exercises in this chapter. We will therefore concentrate on how you can use Fathom to help you understand conceptually what is going on in a hypothesis test. As usual, this will involve simulation.

11.1 Significance Tests: The Basics

Here is the big idea: Significance tests compare reality to the *null hypothesis*. You want to see how easily the results you got could arise by chance from some neutral, usually uninteresting condition. So we define a statistic that measures how unusual our data are (we might use a difference of means, or a difference of proportions, or something more elaborate such as z or t) and compare that to the null, dull, neutral state.

To compare our data with the null, we compare our *test statistic* with a sampling distribution of the same statistic, generated under the conditions of the null hypothesis. Then we see how unusual our test statistic is; the so-called P-value is the probability that a statistic *at least as extreme* as the test statistic could arise from the null condition.

The textbook uses what you have learned about the Central Limit Theorem and about Normal distributions to figure out the sampling distribution; with Fathom, we can do that, but we can also simulate. Some people find that the simulation version helps them understand the logic of the more theoretical reasoning. All of that is a lot to take in; let's do an example in detail.

Example 11.2 Call the paramedics!

Last year, the paramedics responded in an average of 6.7 minutes, with a standard deviation of 2.0 minutes. We assume that the distribution of response times is roughly Normal.

This year, in an SRS of 400 calls, the mean time was 6.48 minutes. The mean time went down. But wait: If you took 400 random calls from the last year (which was $N(6.7, 2.0)$), you *could* get a mean that small, or even smaller. How likely is that?

Let's simulate:

- ❖ In a new Fathom document, make a table with one attribute, **time**. Give it 400 cases, and give **time** the formula **randomNormal(6.7, 2)**. This is a sample of 400 from last year's population.
- ❖ Give that collection a measure called **meanTime**, with the formula **mean(time)**.
- ❖ Look at the value. Rerandomize to make sure the values change and are reasonable.

❖ Collect measures, at least 100 of them. In the illustration, we collected 1000:

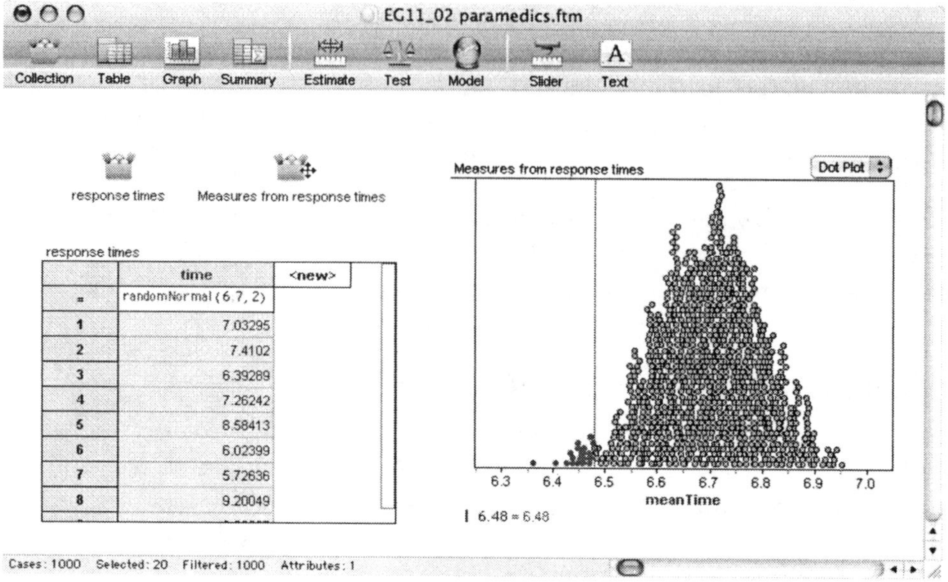

❖ Do a **Plot Value** on the graph to show where 6.48 is. In the illustration, we selected the cases below 6.48 and pointed at the Measures collection. You can see in the status bar that only 20 of the 1000 cases are in the tail of that distribution. That's 2%.

This means that if the true mean were 6.7, the chance that you would get a mean as low as 6.48 is only 2%; it looks like the apparent decrease in mean response time is "real." It doesn't look as if it's due to chance (in particular, sampling variation).

That's what hypothesis testing is all about. You can think of it as a five-step process:

1. We defined the null hypothesis (the distribution of response times is the same as last year's).
2. We defined a statistic that measures what we're interested in (the mean response time in an SRS of 400 responses).
3. Pretending that the null hypothesis was true, we simulated a sample (400 randomly-generated response times, using last year's mean and standard deviation).
4. We generated a sampling distribution of our statistic (many means of 400 responses)
5. We compared our test statistic (this year's mean, 6.48) to the distribution to see how unusual it was. And it *was* unusual: Only 2% of the time, under the null hypothesis, would you get an answer that extreme. Notice that the null hypothesis is about a distribution, while the test statistic is a single number.

One more thing you should keep straight about that value of 6.48 minutes: Suppose the null hypothesis is really, truly true. Would you be surprised if the paramedics took less than 6.48 minutes to respond? No. The next illustration shows a distribution of 400 response times where the mean is 6.7 (last year's figure) and the standard deviation is 2.

Tests and the Sampling Distribution

Most of the time you won't do all five of the steps on the previous page. Remember how the textbook used the Central Limit Theorem and Normal approximations to help with confidence intervals? The same thing happens with these tests. If you know that the sampling distribution is Normal or t-shaped, you don't have to generate it every time—you can just use a formula (one that generally involves the square root of n).

Still, thinking clearly about the sampling distribution will help, both to understand what's going on and to rescue you when various conditions and prerequisites are not met.

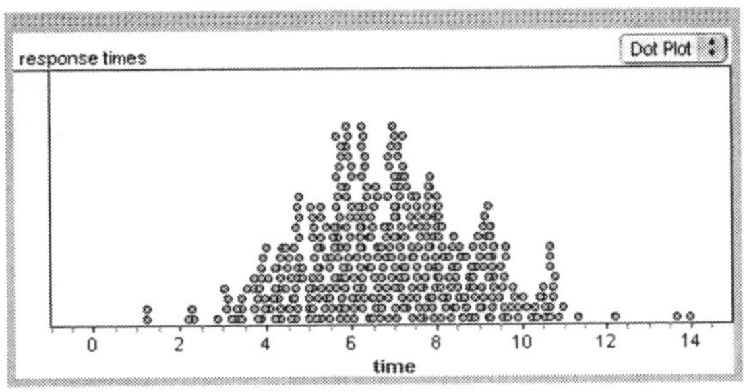

The 6.48 value is practically in the middle! *Lots* of responses are less than 6.48 minutes. But that's not what the test is all about. The test is saying that if you took 400 response times *and averaged them*, then you would be surprised if you got a *mean* that was less than 6.48.

Thus the null hypothesis is really about *two* distributions: the distribution of values (the response times) and also the *sampling* distribution of the statistics. Just to make the point again, here are the two distributions with the same scale on the time axis:

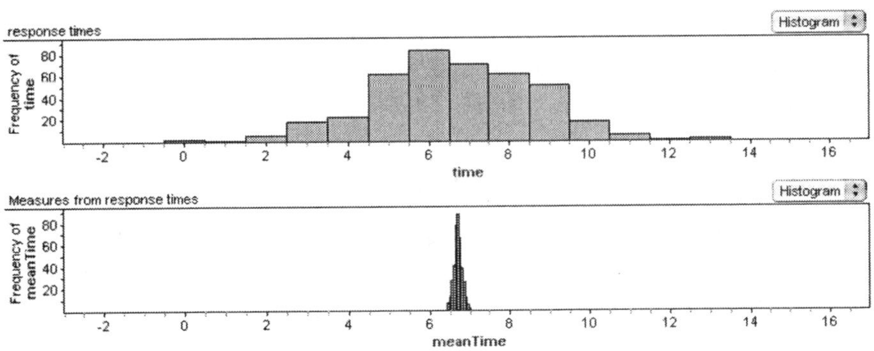

The sampling distribution—the one on the bottom—is much thinner, because the large sample size makes the mean vary a lot less than the values do. How much less? About a factor of 20. Why 20? Because n is 400 and the width of a sampling distribution varies as $\sqrt{n}$.

HOW MANY MEASURES DO YOU NEED? Note that when you create these simulations, the number of cases in the sample (400, in this case) does not have to be the same as the number of measures in the sampling distribution (we used 1000). The sampling distribution is really an abstraction; you can collect as many as you like.

Exercises

In these exercises, make simulations and compare the results to what you get in the textbook.

11.1 STUDENT ATTITUDES. In (b), create the sampling distribution by collecting measures from a collection of 25 values you have defined using **randomNormal( )**. What goes inside the parentheses? In (c), use **Plot Value** to mark the values on your graph. Notice that this exercise doesn't require you to calculate probabilities (though you can).

11.2 ANEMIA I. Again, in (b) simulate to create the sampling distribution. In (c), mark the two values with **Plot Value**. Remember: You need a collection of 50 values and then a sampling distribution (of the mean) with lots of cases, at least 100.

11.2 Carrying Out Significance Tests

Here you can use simulation to get a probability (a *P*-value) empirically. Compare your results with the ones you get by doing an orthodox *z*-test. Remember, simulated results are approximations and vary. You can also get more stable and precise results by collecting more cases in your sampling distributions.

11.27 WATER QUALITY. How many cases should be in your first table? What does each case represent? What variable(s) should the table have? What measure should you collect? After you collect lots of measures, calculate the empirical probability that, if the null hypothesis were true, the dissolved oxygen value would be 4.62 mg/L or less. What does that actually mean in the context of the problem?

11.28 IMPROVING YOUR SAT SCORE. The data are in **EX11_28 Improving SAT.ftm**. You will need the data in order to answer parts (a) and (b) and to calculate the test statistic. This collection is *different* from the one you might use to simulate the null hypothesis.

11.29 PRESSING PILLS. Use file **EX11_29 Pills.ftm**. In (a), another way to assess whether these data could have come from a Normal population is to make a Normal population and simulate drawing 20 cases from it. How often are the samples as non-Normal as our test data?

Doing a z-test in Fathom

Creating the sampling distribution of the mean from scratch is great, and it's good to compare those results with the *z*-test you have learned about in the textbook. But suppose you just want to do the *z*-test? How would you do that?

As of this edition, Fathom's "test" object—you can see it on the shelf—does a *t*-test, which is more appropriate than the *z* under most circumstances. As you saw in the last chapter, the *t* and z distributions are almost identical when n is large, so in many cases, it makes no difference. The textbook introduces the t-test for means in the next chapter. In the meantime, let's do the test with z.

This will seem like overkill, but the technique is useful. The broad question is, how can you use Fathom as a really elaborate calculator? Let's do the next exercise using a z procedure.

11.30 FILLING COLA BOTTLES. Let's say (in advance, as we're supposed to do) that we'll use a significance level of $\alpha = 0.05$ as convincing evidence of underfilling.

- ❖ Enter the data into a collection, naming the variable **volume**. Graph the data and look for outliers. You can see that the data points are fairly uniform and have a mean of about 299—a little low.
- ❖ Make a summary table and drag **volume** to it. You'll see the mean.
- ❖ Choose **Add Basic Statistics** from the table's context menu. Now you can see, among other things, that the standard error is about 0.6.

At this point, you might immediately think, "The mean of this sample is about one-and-a-half standard errors away from 300, and we need about two standard errors for a 5% result, so there is not convincing evidence."

Your instincts are good, but there are two problems. First, we suspect in advance that the bottler is *under*filling the bottles. That means we should use a one-sided procedure. This requires fewer standard errors to get 5%. Second, the standard error in the table is computed from the standard deviation of this sample. That's appropriate for a t procedure, but here we're told that the standard deviation is 3 mL, so the standard error is $3/\sqrt{6}$, or about 1.2. That *reduces* the number of standard errors our sample is away from 300 mL, increasing the corresponding P value.

To compute the right number of standard errors, we'll need $(mean - 300)/(3/\sqrt{6})$.

- ❖ Choose **Add Formula** from the summary table's context menu and enter an appropriate formula, such as **(299.033 – 300)/(3/sqrt(6))**, or more generally, **(mean( ) – 300)/(3/sqrt(count( )))**. Now you'll see the z statistic: the distance, in standard errors, between our sample mean and the "target" mean of 300.

We get –0.789. Since its magnitude is under 0.8, we know immediately that such a mean can arise easily by chance (the lower tail of the Normal distribution beyond $z = -1$ includes over 15% of the distribution, after all), so we could stop now. But suppose we wanted to know the corresponding P value?

- ❖ Make another formula in the summary table. This time, enter **(normalCumulative(–0.789))**. This yields about 0.215, our P-value.

Let's review why that formula works. We want to know the probability that a random result will be at least as extreme as our sample. The sample mean is 0.789 standard *errors* out, which means that it's 0.789 standard *deviations* out in the sampling distribution. The **normalCumulative** function defaults to $\mu = 0$ and $\sigma = 1$, so we need only the –0.789 to get the area of the "tail" of the distribution.

Two questions to think about:

- You could have gotten the same *P*-value using **normalCumulative(1 – 0.789)**. Why does that work? Justify it using a sketch of the Normal distribution.

- If this were a two-sided situation, what formula would you use to compute *P*?

Instead of using a summary table, you could have used computed sliders or even a separate data table to do your calculations. Use any of these techniques on the other exercises in this chapter. But don't worry too much about them; in reality, as with confidence intervals, you will rarely know the population's standard deviation, so you will use the *t*-procedures—and Fathom's "Test" object—which we will introduce in the next chapter.

11.3 Use and Abuse of Tests

Suppose you do lots of tests. Sooner or later, you'll get something that's significant. For example, you'll get a significant result at the 5% level, on the average, one time in twenty *even if the null hypothesis is true!* We can use some of the techniques we just described to illustrate this. (This is related to Example 11.17, *Cell phones and brain cancer* on page 720.)

❖ Open the file **11 Too many tests.ftm**. It looks like the illustration.

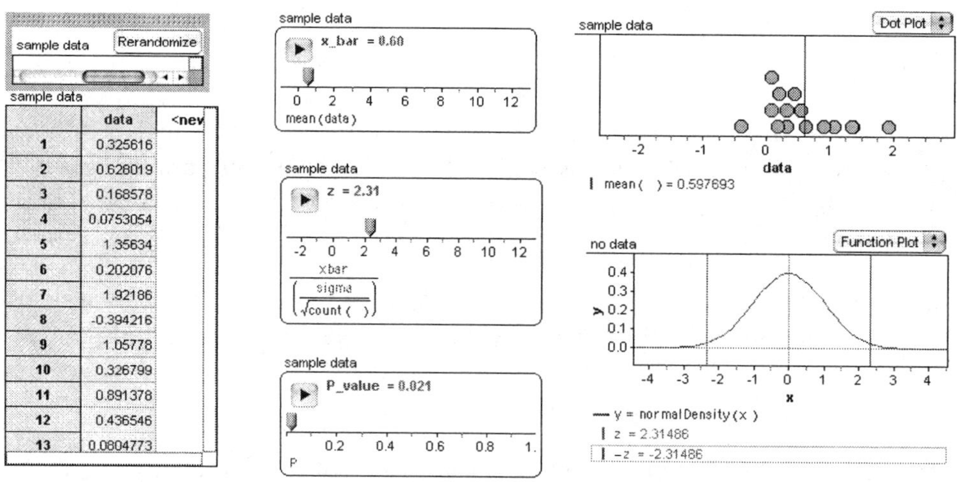

The data are 15 cases randomly generated from a standard Normal ($N(0, 1)$) distribution. The computed sliders show the mean of the sample, its z-value given $\sigma = 1$, and the corresponding (2-sided) *P*-value. The graph in the upper right shows the sample **data**.

❖ Click **Rerandomize** repeatedly. The sample changes and everything updates. Observe how **x_bar**, **z**, and **P_value** change. Notice how the samples themselves look for different **P_value**s. If you get a particularly unusual sample (e.g., $P < 0.05$), take particular note of what the sample looks like.

This is the big lesson: The sample looks as if it probably came from a distribution where the mean is not zero, but it didn't—it's from a standard Normal. If you take enough samples, you'll get something unusual.

Additional exploration

Sometimes you get an unusual observation right away. Sometimes you have to wait quite a while. What happens if we do it many times? We have set up the file with appropriate measures for collection.

❖ Right-click the **sample data** collection (upper left) and choose **Collect Measures** from the context menu.

❖ Collect a lot of measures, at least 100.

❖ Make a plot of **P** and see its distribution. What do you notice? Why does that distribution make sense?

❖ Now plot **xbar** against **P** to see how the sample mean is related to the *P*-value.

Figure out what those graphs really mean. For example, figure out how to use the second graph to determine critical values for **xbar**.

Even more exploration

Take some time to study the formulas in the sliders and in the measures to see how they're set up. See if you can figure out why the formula for **P** works.

Here's a piece of Fathom advice: It's a bad idea to have two variables with the same name. Variables include not only the regular column headings, but also slider names and measures. That's why the measure is named **P** and the slider is named **P_value**.

11.4 Using Inference to Make Decisions

In the last section, we have really been exploring Type I errors: what happens when we reject H_0 when it is actually true. That's relatively straightforward.

Type II errors are harder to understand. The simple explanation is that they happen when you fail to reject H_0 even though it is false. (Notice how we say "*fail to reject*" instead of "*accept* H_0.") But the simple explanation is not enough. After all, if the null hypothesis is only a teensy bit off, we probably won't reject it—and that won't be a problem. It's only a problem if our test fails to detect some genuinely important difference.

Example 11.22 Awful accidents (one last time)

We'll use Fathom to simulate what happens in our tests for various possible "truths." Recall that we're testing our sample of 400 response calls against last year's population of calls which had a mean of 6.7 and a standard deviation of 2.0.

❖ Open the file **EG11_22 Awful accidents.ftm**. This file is based on the previous example in this supplement and works the same way except that this is a one-sided test and we're using the problem's values instead of just a standard Normal distribution.

We also have the slider **trueMu**, which controls the *true* mean of the response time:

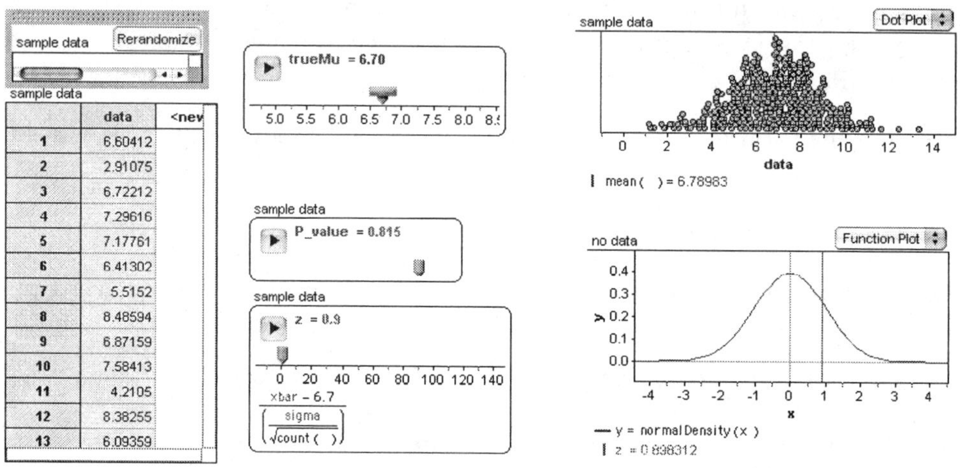

❖ Set the **trueMu** slider to **6.7** (i.e., no change from last year) and press **Rerandomize** a few times. You should see a variety of *P*- and *z*-values corresponding to the distributions of data, just as when we were discussing how you get significant results if you do enough tests (remember, these are Type I errors).

❖ Now set **trueMu** to **6.2**, that is, the paramedics are responding 30 seconds faster than last year on the average. Press **Rerandomize** a few times. Notice that **P** is really small; it probably never even registers, showing only **P_value = 0.000**. (You can find its precise value in the collection's inspector; it's a measure.)

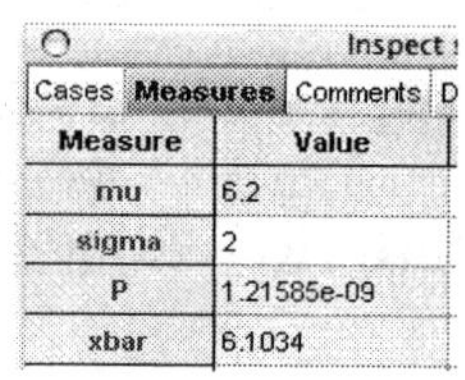

❖ Set the slider to **6.6**. Press **Rerandomize** repeatedly. You'll see that sometimes the **P_value** is less than 0.05 and sometimes it's more.

When it's more than 0.05, we would fail to reject the null hypothesis, saying, "This could be from a population where $\mu = 6.7$." But we would be wrong; *this is a Type II error*.

❖ Set the slider to **6.5**. Now when you rerandomize, you almost always reject H_0. The probability of the Type II error is less.

❖ Set the slider to **6.69**. Now you seldom reject. This is no surprise, since **trueMu** is so close to the null value of 6.70.

Thus the probability of a Type II error depends on the true value of the parameter, *which we never really know*. Nevertheless, when you're planning a study, you can say to yourself, "I think the paramedics could improve by 6 seconds, or 0.1 minutes." Then you can see, in advance, the probability of detecting that change (i.e., rejecting the null hypothesis in favor of the alternative, $H_a : \mu < 6.7$) for our particular sample size.

Power

We can use Fathom—and this very file—to calculate *power* for this situation.

❖ Set **trueMu** to **6.6**.

❖ Collect at least 100 measures from the **sample data** collection.

❖ Plot the distribution of **P** from the **Measures from sample data** collection. Display it as a histogram. Each bin should be 0.05 wide.

❖ Select the bottom bin—the one where $P < 0.05$, as in the illustration. You can even point at it and see how many cases are in the status bar:

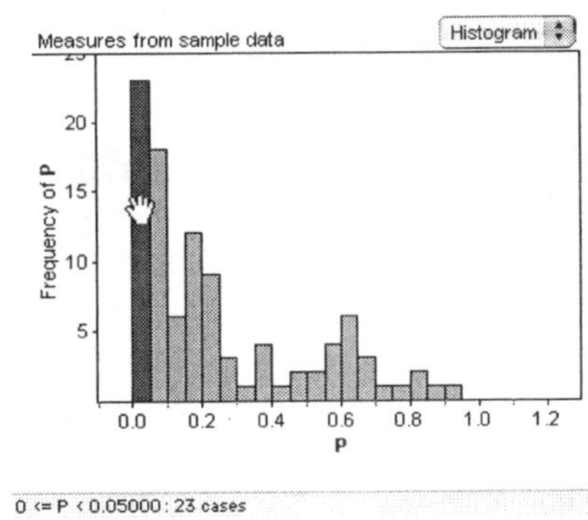

This time, 23 of our 100 cases have $P < 0.05$. That is: 23% of the time, we (correctly) rejected the null hypothesis. In the other 77 cases, we had a Type II error, failing to reject.

This means that when **trueMu** = 6.6, the power is 0.23.

Here are some questions to explore:

✐ What do you get for power if you do it again? (It shouldn't be the same, since we're simulating.)

✐ What could you do in the simulation to get a more precise value for power?

✐ What is the power (approximately) when **trueMu** is 6.5?

✐ For what value of **trueMu** is the power clearly 100%? What does that mean in practical terms?

✐ For what value of **trueMu** is power roughly 80%?

SPEEDING UP THE SIMULATION. If you get impatient waiting for Fathom to collect measures, make its life easier. First, be sure to turn off animation. In addition, close up or delete the graph of the sample, the graph of the density function with its *z*-value, and the table of **data** values.

A power tool

❖ Open the file **11 Power Explorer.ftm**. This file streamlines what we have just done. It looks like this:

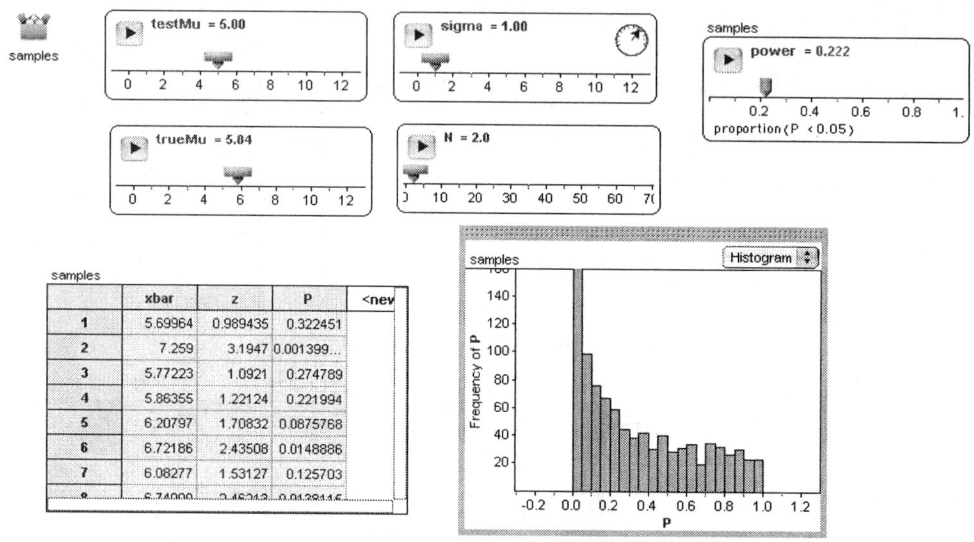

❖ Explore changing **TrueMu** and see how the distribution of **P** changes. Again, the left-hand bin in the histogram is $P < 0.05$.

We would reject the null hypothesis for each of the runs in the leftmost bin. In this case, the null hypothesis is that **trueMu** = **testMu**, which is *false*. The power, then, is the proportion of cases that are in that bin. You will see the **power** (computed) slider change as you change **TrueMu.**

❖ Use this tool to answer the questions in the previous section.

❖ Now explore the effects of **n**. For example, if **testMu** is 5.0, and **sigma** is 1, and you want to be able to detect a **trueMu** of 5.1, how big a sample **n** do you need in order to have an 80% chance of success? (Answer: about 800.)

WHY IS IT SO FAST? Pretty slick, huh? It's fast because we're not collecting measures. The collection, now called **samples**, only simulates sampling, computing **xbar** using **randomNormal** (and the Central Limit Theorem) instead of finding the mean of a sample. So Fathom has a lot less work to do.

12 Significance Tests in Practice

In the previous chapter, we saw how to construct a z-test for the mean. But that assumed that we knew σ, the population standard deviation. In this chapter, we use a more realistic strategy: We use the standard deviation of the sample (s) to estimate the standard deviation of the population. As with confidence intervals, the statistic we make that way, t, does not have a Normal distribution, but rather a t-distribution.

12.1 Tests about a Population Mean

Fathom's built-in test for the mean is a t-test. Tests behave like estimates. Let's see how to use one in an example.

Example 12.2 Sweet cola

❖ Open the file **EG12_02 Sweetness.ftm**. The data are already entered.

❖ Make graphs of **SweetLoss** and assess the distribution for Normality. You should make a Normal quantile plot as shown in the text and find that the mean of the data is +1.0, so there is a loss of sweetness in the sample; is it plausible that this loss is just by chance?

❖ Drag a new test off the shelf. First, change the test to **Test Mean**, then drag **SweetLoss** to the spot at the top of the test where it says **Attribute (numeric): unassigned**.

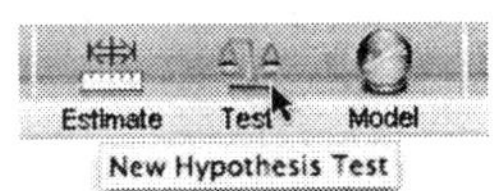

The test fills with calculations and you can read the results of the t-test.

At the bottom of the test, you can see the P-value, which is 0.025. This is a two-tailed test by default, but we believed ahead of time that the sweetness might decrease (giving a **SweetLoss** greater than zero). Therefore, you can use a one-sided test. How do you do that?

❖ In the test itself, find the place where it says **SweetLoss is not equal to 0**. That phrase is the embodiment of two-sidedness. Click on **is not equal to**; it turns into a menu. Our alternative hypothesis is that **SweetLoss** is greater than zero; choose that.

The test changes, some additional text adjusts, and the P-value gets cut in half to 0.012.

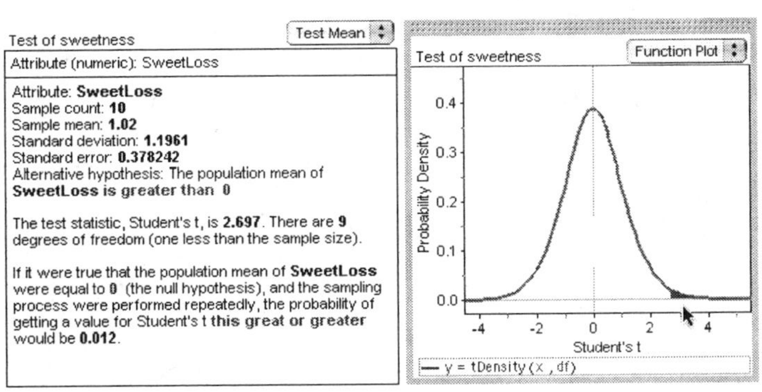

- ❖ If you want to see the distribution that applies, right-click on the test and choose **Show Test Statistic Distribution** from the context menu.

- ❖ Explore the test and the graph: Change it to **less than 0** and see what happens. Change the number as well, for example, test whether **SweetLoss** is greater than 0.5. Is the change still significant?

As with the estimate, if you don't want all the text, you can un-check **Verbose** in the context menu. Beware, however that the text in Fathom's tests is worded carefully to represent what the test is really telling you. If your instructor asks, "What does that *P*-value 0.012 mean?" and you answer, "It means there's a 1.2% probability that we're wrong, and the true change in sweetness is actually negative," the angels (or at least your instructor) will weep.

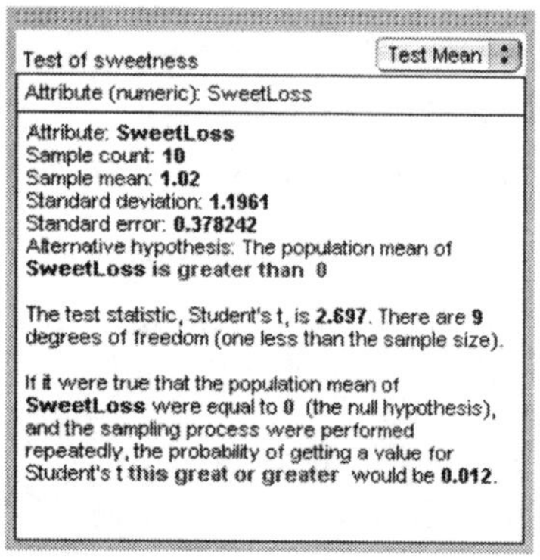

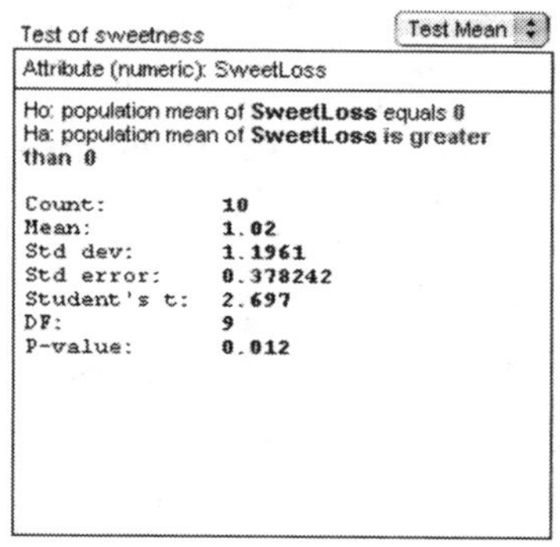

Example 12.3 Diversify or be sued!

- ❖ Open the file **EG12_03 Portfolio.ftm** and do the usual exploratory dance. The illustration shows a histogram of the **return** data and a Normal quantile plot. You can see that the 39 data points are more or less Normal and that the mean is clearly less than zero, so it's certainly less than 0.95%, which is the population parameter. Is it significant, though?

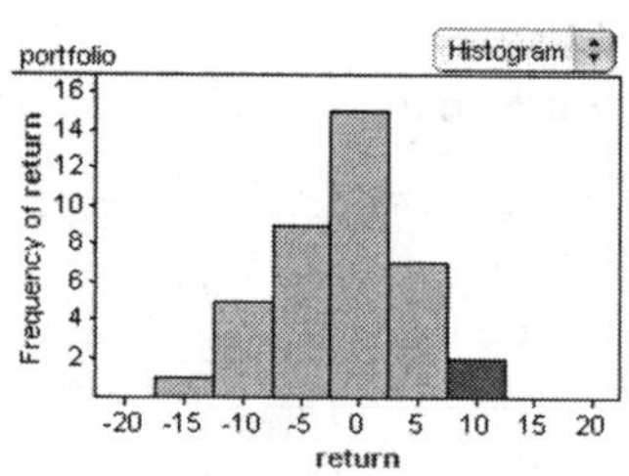

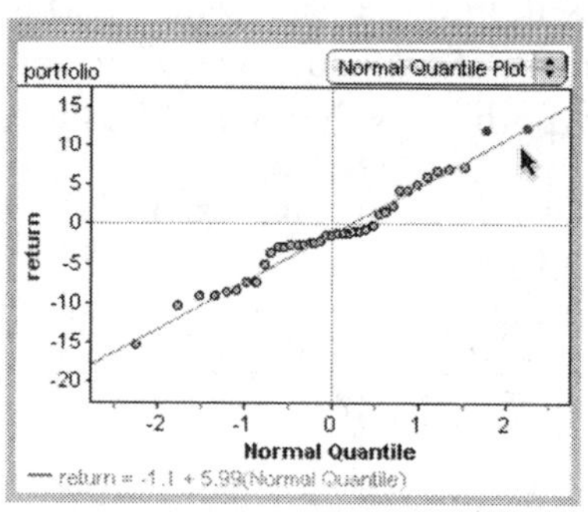

- ❖ Make a new test. Choose **Test Mean** from its corner menu and drag **return** to the correct spot.

- ❖ Edit the **0** in the test (to which the mean of return is being compared) and change it to **0.95** (the assumed population parameter). The test is already two-tailed.

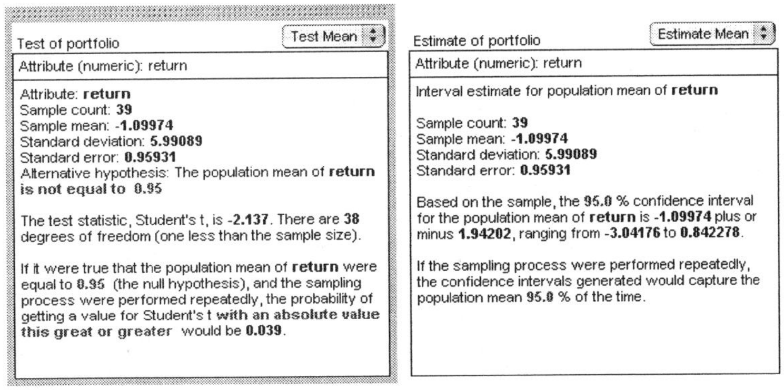

The illustration shows the test (on the left) and the *P*-value of 0.039. On the right, we see a 95% confidence interval (a Fathom "estimate") for return; you can see that 0.95 is not in the confidence interval.

Exercises

12.5 Healthy bones. The file **EX12_05 Healthy bones.ftm** has the data. Be sure to make a graph or two before you make the test; this will help you select the outliers to **Cut** for part (b).

12.6 Heat through the glass. Use file **EX12_06 Heat glass.ftm**. This exercise asks for an interval estimate (i.e., a confidence interval). Does the confidence interval also give you "evidence that the conductivity…is greater than 1?"

12.7 Eating and weight gain. Use file **EX12_07 Eat and gain.ftm**. In this file, we have already done the conversion to pounds. Notice how this setup is a paired design.

12.8 NEAT. In this exercise, you'll see how to do the *t*-test by entering summary data instead of raw, case-by-case data. You've already learned how to do this for confidence intervals; it's pretty much the same procedure.

Notice how the problem statement gives you the relevant quantities: There are 16 subjects, the increase was 328 calories per day, and the standard deviation in that increase was 256. It also tells you to test the null hypothesis that there was no change in NEAT; that is, H_0 says that the increase is zero. The text doesn't say so, but let's look for $P < 0.05$ (so it will correspond with the 95% confidence interval in (b)).

- In a new document, drag a Test off the shelf. Change it to **Test Mean**.
- Edit the blue values. Change **AttributeName** to **increase** and so forth. Your test should end up looking like the illustration. Notice that we did not have to change the 0 or the "not equal to," since we're doing a two-sided test.

From Summary Statistics   Test Mean

Attribute (numeric): unassigned

Attribute: **increase**
Sample count: **16**
Sample mean: **328**
Standard deviation: **256**
Standard error: **64**
Alternative hypothesis: The population mean of **increase is not equal to 0**

The test statistic, Student's t, is **5.125**. There are **15** degrees of freedom (one less than the sample size).

If it were true that the population mean of **increase** were equal to **0** (the null hypothesis), and the sampling process were performed repeatedly, the probability of getting a value for Student's t **with an absolute value this great or greater** would be **0.00012**.

- ❖ For part (b), construct a confidence interval (a Fathom estimate), entering data the same way.

Example 12.5 Floral scents and learning: Performing a paired t test

You can do tests on paired data using the same strategy we used with confidence intervals. You'll make a new variable, **diff**, which should be the difference in times. Use the file **EG12_05 Floral learning.ftm** for the data. You'll get the disappointing result that we cannot claim that smell affects your learning. Too bad!

Example 12.6 Diversify or be sued II

In this example, we return to the issues of power and Type II errors. As the textbook suggests, we can use our "z" understanding of power and assume that we know σ, in order to assess the power of a test.

You can use **11 Power Explorer.ftm** to do this.

More exercises

12.9 RIGHT VERSUS LEFT, I. Use the file **12_09 Right vs left.ftm** for the data. Be sure to create a new variable, since you're doing a paired test. Also, it's good to have the "positive" (i.e., hoped-for) result correspond to a positive value, so make sure that your variable is positive if the subject did better on the right thread than on the left.

In part (d) you can use the file **11 Power ONE TAIL.ftm** to investigate the power of a one-tailed test in the positive direction (i.e., the alternative hypothesis is that the population mean of the variable in question is positive).

12.11 MUTUAL FUNDS PERFORMANCE. In this exercise, you enter the summary data into a test. We don't have raw data.

12.18 GROWING TOMATOES. Enter the summary data here as well.

12.19 THE POWER OF TOMATOES. Use one of the power documents we have referred to (they're in the Chapter 11 folder). You'll need to figure out which one is suitable. Is this a one-sided or two-sided test?

12.21 VITAMIN C CONTENT. Use **EX12_21 Vitamin C.ftm** for the data.

Fathom and the t

You can use Fathom to make *t* calculations just as we have made Normal calculations ever since Chapter 2. Fathom's *t*-distribution functions are completely analogous to the Normal functions. You will find them under **Student's t** in the formula browser. The only difference is that they require one more parameter, the number of degrees of freedom, which is usually one less than the sample size.

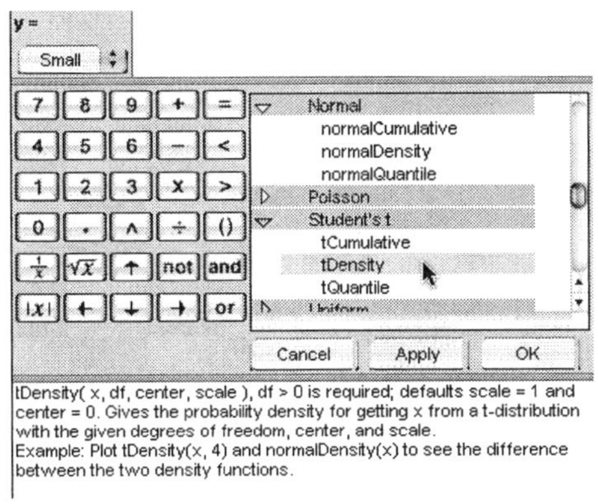

12.2 Tests about a Population Proportion

In Chapter 10, you learned about confidence intervals for means and then, using the Normal approximation, confidence intervals for proportions. Then in Chapter 11, you did hypothesis tests for means and talked about how they are kind of parallel to the confidence intervals. It should come as no surprise, then, that we want to do hypothesis tests for proportions. These tests will be related to everything else.

	Confidence Intervals a.k.a. Interval Estimates	Hypothesis Tests
Means	*z* procedure *t* procedure	*z* procedure *t* procedure
Proportions	*z* procedure (Normal approximation)	You Are Here

As with other hypothesis tests, we'll set up a null hypothesis (e.g., "50% of the population favors Proposition 1138"). Then we'll try to show that it is implausible in the face of an alternative hypothesis (e.g., "Less than 50% of the population supports Prop 1138."). To do that, we invent a statistic and compare it to the sampling distribution produced under the conditions of the null hypothesis. We try to show that our test statistic is so far out on the tail of the distribution that we are confident the null hypothesis is false.

Be sure you've read this section in the textbook. It begins by building on previous work and on the Normal approximation, and introduces a one-proportion *z* test that is completely analogous to the one we used for means in Chapter 11.

We'll take a slightly different path here, returning to first principles by simulating the situation in Fathom. You're practically an expert by now at making simulations; we'll make a sample using random functions and then collect measures to create our sampling distribution.

Example 12.7 Work stress

We'll use an example from the textbook:

❖ Make a new Fathom document and a new table with one column, **stress**. Give the table 100 cases. Why 100? Because that's the size of the sample of workers in the large restaurant chain. Name the collection **sample**.

❖ Give **stress** the formula **random( ) < 0.75**. Use **random( )** to make a random number between 0 and 1; use 0.75 because the problem states that 75% of workers nationally feel that stress has a negative impact on their personal lives. (These people will have **stress** = **true**.) The collection, then, is an SRS from the national population under the conditions of the null hypothesis, namely $p = 0.75$.

❖ Now you need a statistic. Use the proportion of people in the sample who are stressed. Make a measure (it's in the **Measures** panel of the inspector) called **propStressed** and give it the formula **proportion(stress)**.

USING BOOLEANS AGAIN. We haven't used these Boolean attributes in a while. That formula, **propStressed = proportion(stress)**, is equivalent to using a wordier formula such as $propStressed = \dfrac{\text{count}(stress = true)}{\text{count}()}$.

At this point, your document should look something like the illustration. Notice how the measure's value is 0.77; it's not exactly 0.75 because **sample** is a random sample where each element has a 75% chance of being true.

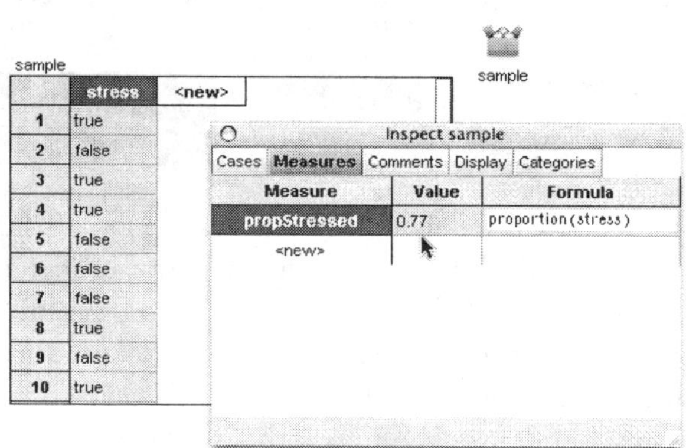

Now, is the 68/100 poll result unusual? To answer that question, **Rerandomize** a few times in order to make sure things are working, then…

❖ **Collect Measures** (it's in the collection's context menu). Collect several hundred.

❖ Graph **propStressed** and use **Plot Value** to add a line at 0.68 (our test statistic). Your graph should resemble the next illustration. There, 26 of the 500 cases, or 5.2%, **propStressed** is as low as 0.68 or lower (your results will vary, of course).

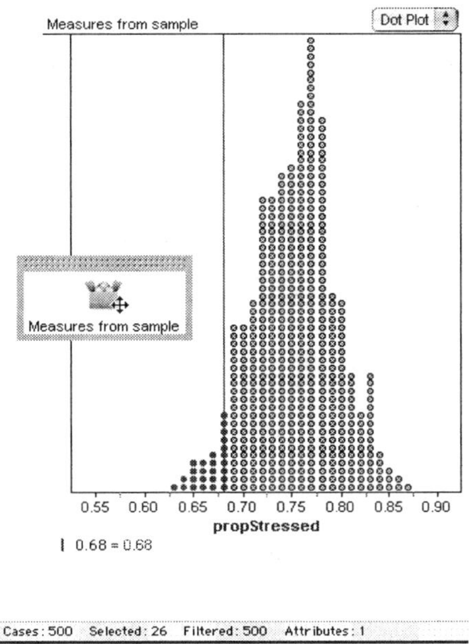

Thus the empirical *P*-value is 0.052; if you run it again, you will get a slightly different value.

How would you use Fathom's test object to do the same thing?

❖ Drag a test off the shelf and choose **Test Proportion** from its corner menu.

❖ Edit the relevant blue text, choose the one-tailed menu text (change **is not equal to** to **is less than**), and show the p-hat distribution (it's in the test's context menu):

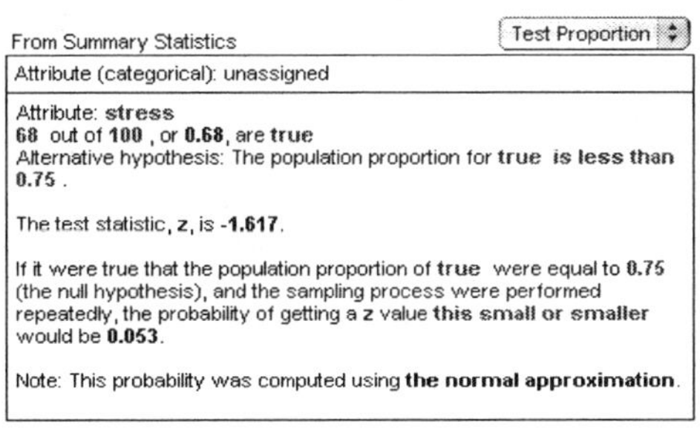

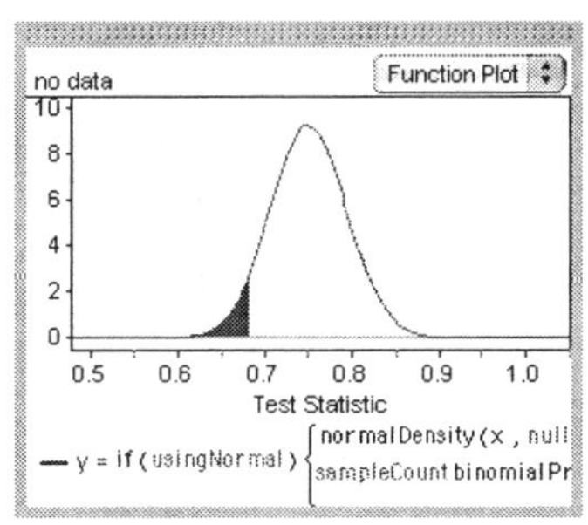

Notice how the computed *P*-value agrees with the empirical one and the distribution resembles the one from your simulation.

Notice also that the test tells you that it used the Normal approximation. What alternative is there? The exact binomial calculation. You would use it when your situation does not meet the various conditions for using the Normal.

When conditions aren't met

What can you do when conditions are not met? It turns out that the logic of the simulation works in every case, because it does not depend on the Normal approximation. Let's look at an exercise to see how this would work:

12.23 CHECKING CONDITIONS. Let's be concrete. Suppose a coin comes up heads 8 out of 10 times. We want to test the hypothesis H_0 that the coin is balanced, i.e., $p = 0.5$.

You cannot use the one-proportion *z*-test (why not?) so you need a different strategy.

- ❖ In a new document, make a new table named **coin** with one attribute, **heads**.
- ❖ Give it ten cases and the formula **random( ) < 0.5**.
- ❖ Make a measure, **nHeads**, with the formula **count(heads)**.
- ❖ Collect at least 100 measures. Plot **nHeads** and use **Plot Value** to plot 8.
- ❖ Count how many samples of ten coins had 8 or more heads.

You should also decide whether to count the runs with 0, 1, or 2 heads. Is this really a two-tailed situation? Put another way: Are you really testing the coin for fairness or for being heads-heavy?

The illustration at left shows what your graph may look like. There are five cases for 8 and 9, and four for 1 and 2, a total of 9 (or 9%) at least as extreme as an 8-headed sample. Unusual, but not at the 5% level. On the right is what you get with the Fathom test. Notice at the bottom—Fathom used the exact binomial to get its *P*-value because the traditional conditions for using the Normal approximation were not met.

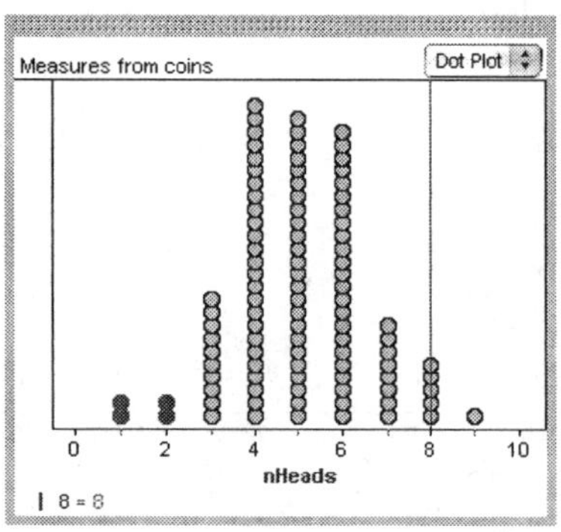

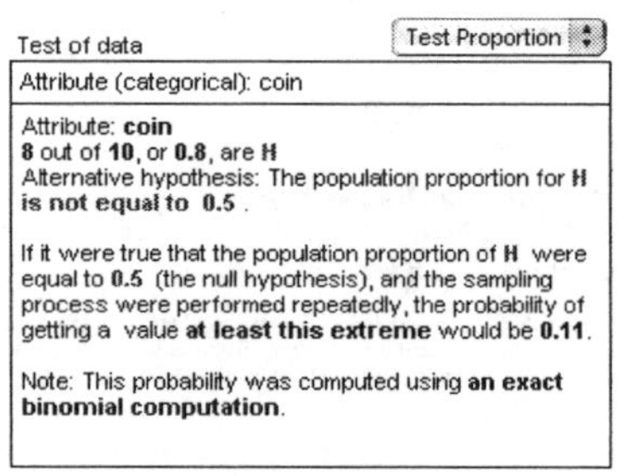

- ❖ Create simulations for part (b) and (especially challenging) part (c).

In part (c), you will have to make a population of 250 students split half and half, and sample *without replacement* in order to simulate the null hypothesis. (Why is that so important?) As a consequence, you will need the source-sample-measures structure (see Chapter 8).

Power and proportion

You can use a power tool, **12 Power in proportion.ftm**, to help you answer power questions for one-sample z-tests of proportion.

Play around with this file to see how it works.

✐ Observe that the one-tailed P-value is always half the two-tailed P-value. Explain why.

✐ Observe that the one-tailed power is always greater than the two-tailed power, but not twice as great. Why not?

Exercises

12.24 We want to be rich. Use a Fathom test to answer this question. Drag a test off the shelf, decide what kind of test you want, and fill in the numbers. For part (b), use an estimate instead.

12.25 Hack-a-Shaq. In part (c), use the power tool file. (You have to decide if it's a one- or two-tailed situation). But what numbers do you put in the sliders? Use 0.533 for **p0**—it's the number we're testing against. **trueP** is given as 0.60. Calculate sigma using the regular formula $\sigma = \sqrt{p(1-p)}$. Use 39 for **n**.

✐ One step further: How many free throws would he have to shoot at 60% to have a 90% chance of rejecting last year's 53.3%?

12.27 Work stress and power. Again, use the power tools. In part (d), you will have to alter the tools to change the significance level. The only formula you have to change is the one for the **power** slider. It will be obvious what to do.

12.29 Risky behavior. Answer the questions as stated, but then check out the file **EX12_29 Risky behavior supp.ftm**. It contains data from a different SRS, part of a national health survey. This sample consists of 359 heterosexual adults who say they have had more than one (opposite-sex) sexual partner in the last year. They also report the number of times they have had sex without a condom in the last 30 days.

✐ What is a reasonable [interval] estimate for the proportion of those U.S. adults who have had more than one sexual partner in the last year, who have had sex without a condom in the last 30 days?

✐ Why is that figure so different from the proportion in the problem in the textbook? Come up with as many explanations as you can.

✐ Does this proportion appear to differ by sex? By education?

❖ You will probably need to create some new columns. Also, to answer the last question, a summary table may help.

The calculations in most of the review exercises are perfect to do in Fathom. Some notes and reminders:

12.32 ATTITUDES TOWARDS NUCLEAR POWER. For this problem, drag a new estimate off the shelf, use its menu to specify what kind of estimate you want, and enter the summary statistics from the problem (225 out of 512).

Use this same basic procedure in 12.33–12.36.

12.37 SHARKS. Use the file **EX12_37 Sharks.ftm**. It has all the data.

12.38 IS THE BELGIAN EURO COIN "FAIR?" This is a great exercise to do with a simulation. In your simulation, use the *number* of heads instead of the *proportion* of heads as your statistic. Because this is a simulation of a specific setting, using the number of heads as a statistic works just as well as the proportion you need in the more orthodox *z*-test.

✐ Why would you *want* to use the number? Are there any advantages?

13 Comparing Two Population Parameters

Until now we have been dealing with one-sample statistics: confidence intervals and hypothesis tests about means or proportions. When we have compared data, we have generally compared a null-hypothesis distribution to our *test statistic*—a single number derived from the data. Even when we did paired analysis (for example, looking at pre-post tests), we first reduced the data to a single sample: a set of data consisting of the *differences* between pre and post.

In this chapter, we compare data from two genuinely independent samples. As before, we'll try to get at the underlying logic through simulation. But this time, instead of producing random numbers using sampling or random-number generators, we'll do something fundamentally different: We'll *scramble* the data.

Even though scrambling is new, its purpose is still to simulate the null hypothesis and compare our test statistic to a sampling distribution we generate when the null is true. When we compare two samples, our null hypothesis will be: *The two samples are from the same population.* The purpose of a hypothesis test is to show that this statement is implausible.

Scrambling, as you will see, is an ingenious way to make the null hypothesis true. You take all the data you have, this set of data from two groups, and *scramble the group assignments.* This makes it so that group assignment has nothing to do with the response variable, so any difference between the groups is due to chance alone. Calculate the statistic, e.g., the difference of means or a difference of proportions. Repeat the process many times to build up a sampling distribution. If the test statistic looks unusual next to that distribution, you can reject the null hypothesis and conclude that the two populations are not the same.

13.1 Comparing Two Means

We'll begin with an example from the textbook. We'll see what happens in simulation. As an added bonus, since our reasoning does not depend on the Normality of the data, the data do not need to meet the Normality condition (page 782) if you use simulation. Afterwards, we'll show you how to use Fathom's built-in tools.

Examples 13.2–13.5 Calcium and blood pressure

❖ The file **EG13_02 Calcium.ftm** has the data already entered. Open it.

❖ Make a graph and a summary table so you can see roughly what the data look like and how the groups differ. You might see this:

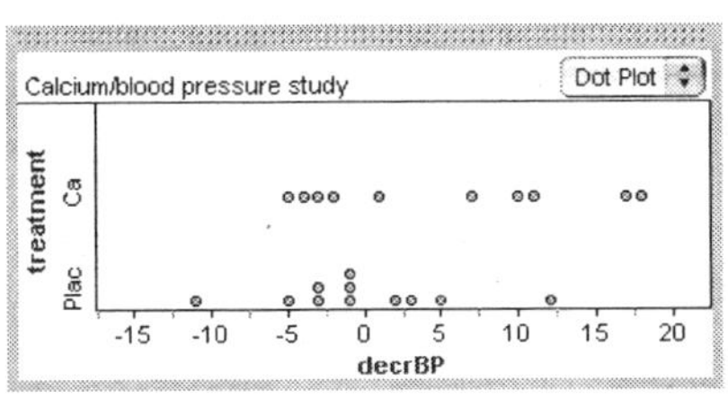

Calcium/blood pressure study

	treatment		Row Summary
	Ca	Plac	
decrBP	5	-0.27272727	2.2380952
	10	11	21
	8.7432514	5.9006933	7.687033

S1 = mean( )
S2 = count( )
S3 = s( )

You can see that **decrBP** (the decrease in blood pressure) is larger, on the average, for those in the **Ca** (Calcium) group compared to the **Plac**ebos. There is a lot of overlap, though. Even if there were no difference between the groups, this much difference between samples would happen occasionally. But how often?

To investigate that question, imagine writing the data on cards, shuffling them, and then dealing them out to the **Ca** and **Plac** groups. We could see how different those random groups are from one another. To do that quantitatively, we need a statistic: a measure of the difference between the groups.

We could use any of a number of measures, but the obvious one is the difference of means. Because we expect Calcium to reduce blood pressure, we'll use this:

mean(decrBP, treatment = "CA") – mean(decrBP, treatment = "Plac")

That way, the measure will be large and positive when the effect we expect is strong.

❖ Enter that formula as a measure called **diff**. Note its value for your data: about 5.27. This is our test statistic.

❖ Now you're ready to generate the sampling distribution of **diff**. Right-click on the collection and choose **Scramble Attribute Values**. A new collection appears. Look at it in a table, and you'll see that all of the data are the same, except that the data in the first column have been scrambled.

❖ Graph the data in the new collection.

This is a collection in which the null hypothesis—that the Calcium and Placebo treatments are the same—is true.

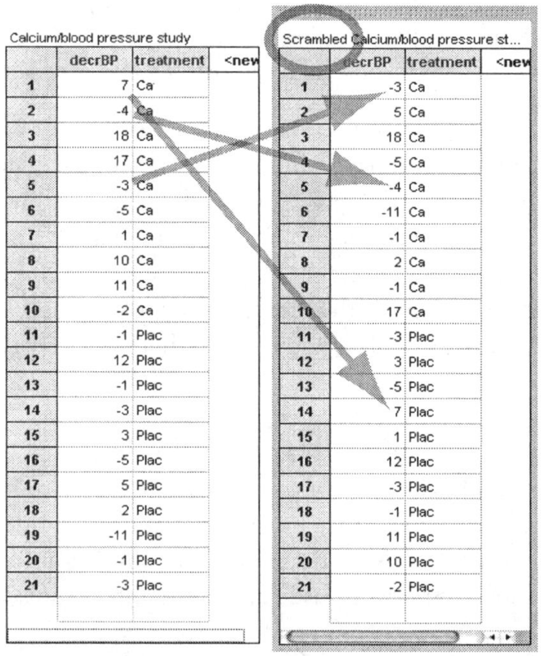

❖ Inspect the Scrambled collection. Go to the **Scramble** panel and press **Scramble Attribute Values Again**. See how the data change.

❖ Inspect the **Scrambled** collection and look at its measures. You'll see that it has a value for **diff** computed from its data; it "inherited" the definition of **diff** from the original data collection. See how it changes when you rescramble.

❖ Right-click on the new collection and choose **Collect Measures**. A third collection appears, a Measures collection. Collect at least 100 measures and plot the distribution.

❖ See where the test statistic, 5.27, lies in this distribution. Does it look unusual? It's off on the edge of the distribution. Only 6 out of 100 values as extreme as your test statistic:

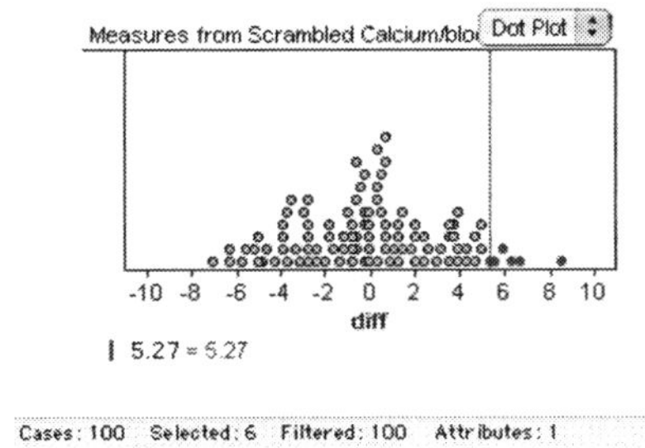

It's not clear whether you should reject the null hypothesis. If we have set $\alpha = 0.05$, we should collect more measures in order to get a more secure value for P.

Some observations:

First, this is a one-sided situation because we expect ahead of time that the Calcium supplements will reduce blood pressure. There are many strategies for simulating the two-sided version of this test, including defining **diff** as the *absolute value* of the difference of means.

Second, you might have thought of entering the data in two columns: **Calcium** and **Placebo**. As we said early on, that can bring you trouble: although it makes sense to us humans, to the computer, the implication is that each row in the table is a case, and that the Calcium and Placebo values in each row are related to one another. Not so.[1]

Third, the textbook goes on to devise two-sample z and t statistics. We used a plain old difference of means, not dividing by standard deviations or $\sqrt{n}$ s. Does what we did still work? You bet. This is a legitimate statistical procedure (a permutation test). This kind of test works with *any* statistic. Suppose we use a *ratio* of means? No problem! Suppose we wanted to test the hypothesis that the *variances* in the two groups were different? Piece of cake! But this test-by-simulation is completely specific to this particular data set. With different data, you have to set it up again: define a measure, find a test statistic, re-scramble, generate the sampling distribution—the whole ball of wax. With the t procedure, you need only a few means and standard deviations, and with those, shazam! you get a P-value.

A tool like Fathom makes rescrambling (or using any of the simulation techniques we have learned) actually practical in a way that it was not, way back in the 20th century. So if you ever get confused about what a P-value means, or doubt the output of some statistical computation, don't hesitate to whip up a simulation and see for yourself.

Using t-procedures in Fathom

As you might expect, Fathom's estimate and test objects will perform t calculations for you. Continuing with the same file,

❖ Drag a new test off the shelf. Choose **Compare Means** from its corner menu.

1 If we had identical twins and chose which one got the Calcium at random, this data structure would make sense, but then *the analysis should be paired,* and we would use one-sample techniques.

❖ Drag **decrBP** (the decrease in blood pressure, the response variable) to the top slot. It reads **First attribute (numeric): unassigned**.

❖ Drag **treatment** to the second slot, the one that reads **Second attribute (numeric or categorical): unassigned**.

Fathom fills in all sorts of information about the situation, including the *P*-value, about 0.13. It's high because it is still a two-sided test. You need to fix that.

❖ Under the alternative hypothesis, use the popup menu to change **is not equal to** to **is greater than**, as shown:

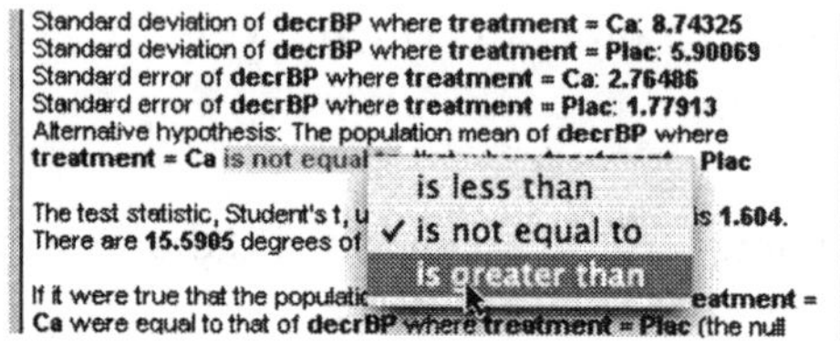

The text readjusts and the *P*-value changes to 0.064, consistent with the one in the textbook.

❖ Follow the same steps with an estimate. Choose **Difference of Means** from the corner menu and drag the same variables in (you can even drag them from the test). Confidence intervals are not one-tailed, so you can either leave it as is or change the confidence level to 90%, so that it leaves 5% on either side of the center. The illustration shows both the test and the estimate with **Verbose** un-checked:

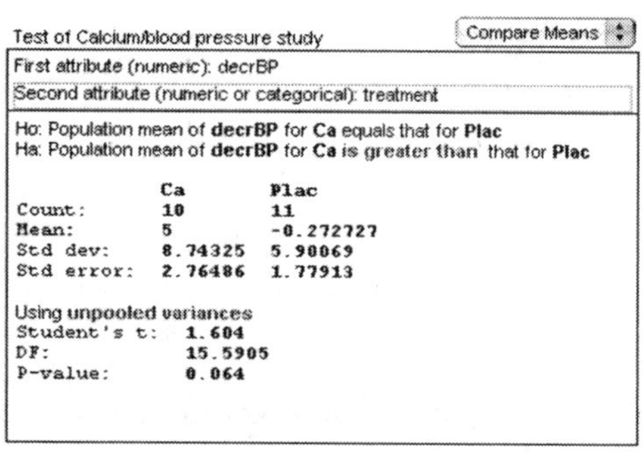

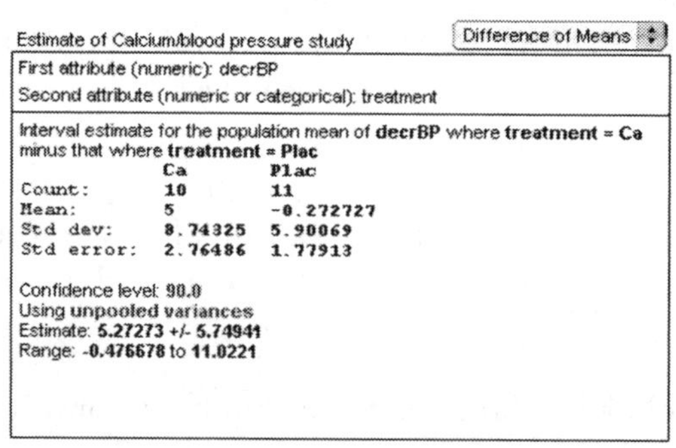

We should explain two mysterious things. The first is that Fathom is using **unpooled variances**. That's good. You could change it to **pooled** by treating the text as a menu, but don't. Your textbook explains why that's a bad idea later in Chapter 13. The second is the non-integer number of degrees of freedom (*df* = 15.5905). This is explained in the section *Software approximation for the degrees of freedom* on page 792. You can compare these results to the results in Example 13.5.

Exercises

You can now use Fathom to help you with many of the exercises in this section. For example, use Fathom in exercises 13.5 and 13.6 to help you graph and visualize. Use Fathom for Exercise 13.12, where you analyze your paper-airplane data. Here are some additional directions:

13.13 Is red wine better than white wine? The file **EX13_13 Wine test.ftm** has the data from the textbook. Use Fathom to do part (b).

13.14 Red wine or white wine? The file **EX13_14 Wine interval.ftm** is the same as the file from Exercise 13.13. Use it to do part (b).

13.15 Competitive rowers, I. As with the other tests and estimates we have used, you can enter summary statistics directly into the test if you do not have the raw data.

- Make a new blank Fathom document and drag in a new test.
- Change it to **Compare Means**.
- Enter the summary statistics from the textbook's SAS output into your test. Make sure you get results consistent with the SAS printout and answer part (a).
- Enter the same numbers into a new *estimate* to answer part (b).
- Answer part (c).

13.16 Competitive rowers, II. Use the same techniques to answer the question in this exercise. Just enter the summary statistics from SAS into the appropriate Fathom object(s).

13.24 The effect of logging. Use the data in the file **EX13_24 Logging.ftm** and Fathom to help you answer the questions.

13.2 Comparing Two Proportions

As with comparing means, we'll start with simulation and scrambling. The whole process is identical, but there are a few things to watch out for.

Example 13.7 Prayer and in-vitro fertilization (IVF)

In the problem statement, we learn how many women successfully got pregnant when prayed for, how many were in that group, how many in the control group got pregnant, and how many were in that group. When we use the tests directly, we'll just edit the blue text to get our answer. To simulate through scrambling, however, we will need to have the raw data. Because there is no easy way (yet) to transform raw counts into a collection, we did it for you.

- Open the file **EG13_07 Prayer and IVF.ftm**. It contains 169 cases, one for every woman in the study. Each case has two attributes: **pregnant** (**true** or **false**) and **group** (**prayer** or **control**).

Begin, as you always should, by making a graph and maybe a table. You may want to review Section 4.2 (page 67) in this guide for a reminder about how to deal with categorical data in graphs and tables. We will go right to the ribbon chart:

- Make a new graph and drag **group** (the predictor variable) to the horizontal axis. You get a bar chart. Change it to a ribbon chart.

❖ Drag **pregnant** (the response variable) to the *middle* of the graph, splitting each segment of ribbon. Your graph should look like this:

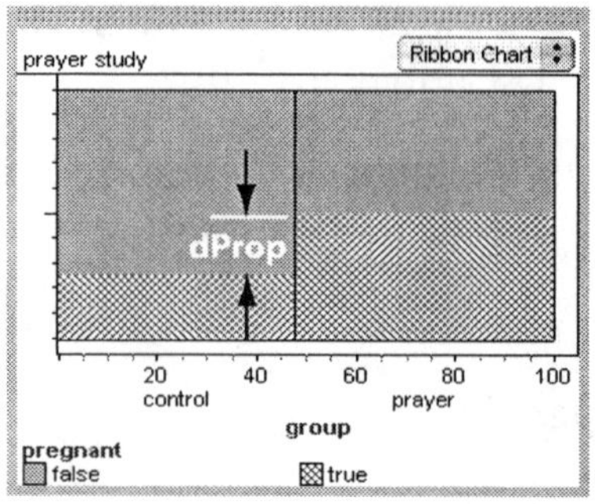

A bigger proportion of the women in the **prayer** group got pregnant. If the proportions had been the same, the two horizontal dividers would meet at a single point. One measure of the difference between the groups is the degree to which those lines miss each other: the length of the vertical segment **dProp** in the graph. We have called it **dProp** because it's the difference in proportions between the two groups. One good formula would be:

proportion(pregnant, group = "prayer") – proportion(pregnant, group = "control")

EFFECTIVE USE OF BOOLEANS. We have once again used a Boolean variable to make writing formulas easier. Because **pregnant** has the values **true** and **false** (instead of "pregnant" and "not pregnant"), the name of the variable itself, **pregnant**, is a Boolean expression. We also use the "implied filter" in the formula so that each **proportion( )** function will find the proportion of just its own group—**prayer** or **control**.

❖ Make a measure in your collection, calling it **dProp**. Use the above formula or any other formula that gives you the same result (about 0.24). That's our test statistic. Under the null hypothesis—that prayer makes no difference—it should be close to zero.

❖ As before, scramble attribute values, then collect measures and construct your sampling distribution. We took 100 measures and got the distribution of **dProp**:

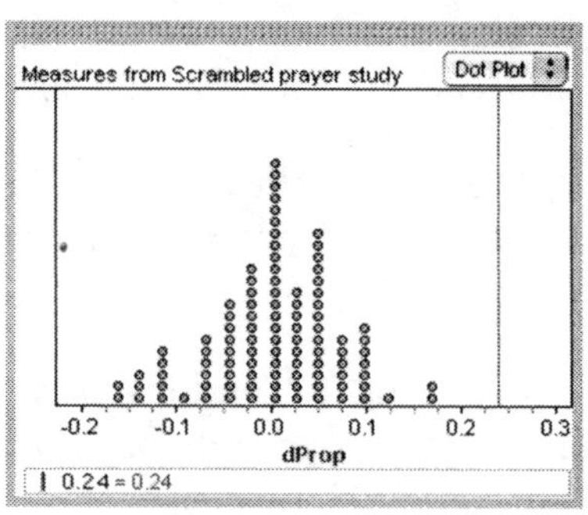

As you can see, *none* of our 100 scramblings were as extreme as our actual data. This is strong evidence against the null hypothesis.

What we have done here is basically the same as what we did for a two-sample difference of means. Schematically, it looks like this:

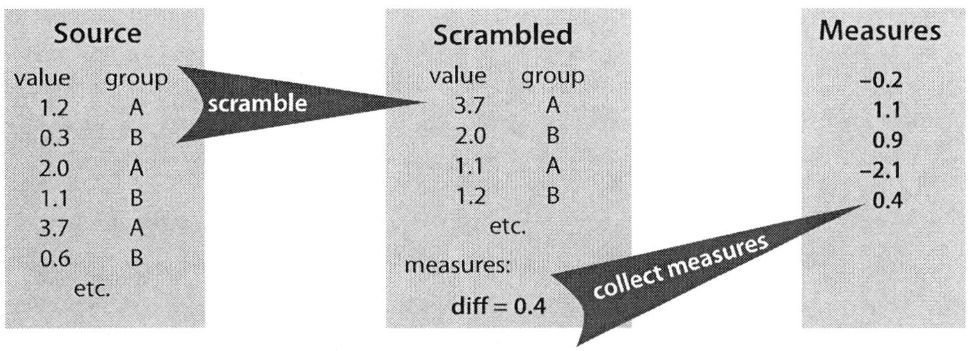

The only tricky parts were generating the raw data and writing the formula for the difference in proportions. The latter is most problematic. Sometimes you will have to write complicated formulas to get what you want. Just go slow and check your work by seeing if the formula is giving you the right numbers. People seem to have the most trouble with:

(a) Remembering that many categorical values need quotes, as in **group = "control"**.

(b) Using **proportion( )** incorrectly. See the gray box on the next page.

(c) Failing to use parentheses when you need them.

(d) Being overwhelmed by a huge formula. If a formula gets too complicated, break up your measure into several measures, e.g., **controlProp** and **prayerProp**. Then your ultimate statistic, **dProp** can have a simple formula: **prayerProp – controlProp**.

Using the Fathom objects

As with the two-sample difference of means, Fathom's test and estimate objects can do the job instantly. In the test, choose **Compare Proportions** from the test's corner menu; in the estimate, choose **Difference of Proportions**.

If we use a test and an estimate on the prayer and IVF data, we get the illustration (again with **Verbose** turned off). In general, you will have to mess around with the blue text to make sure that Fathom is comparing the proportions you particularly want to compare.

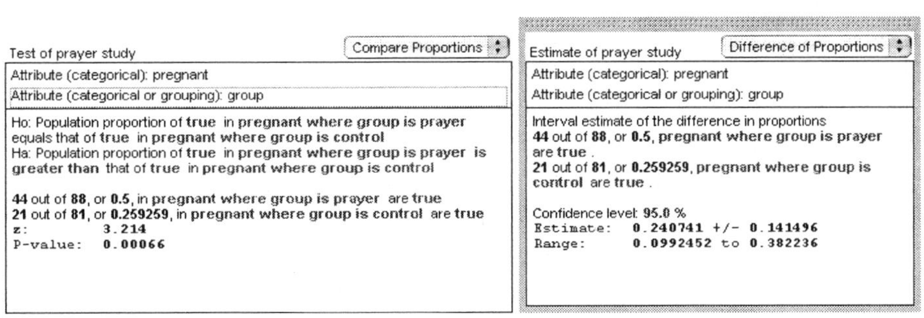

Using the proportion( ) function

Put a Boolean expression inside the parentheses to signify what you want to find the proportion of. This is the numerator if you were to write your proportion out as a fraction. If you want to find that proportion within some subset of the whole collection, put another Boolean that describes the subset after a comma (making an implied filter). That will be the denominator. Some examples might help:

proportion(sex = "Male")	The proportion of males in your collection.
proportion(sex = "Male", age > 17)	The proportion of adults who are male.
proportion(age > 17, sex = "Male")	The proportion of males who are adult.
proportion(thirsty)	The proportion of all people who are thirsty. This only works if **thirsty** is itself a Boolean (true/false) variable.

Nobody can get this right reliably! Therefore, always check your computer results to make sure you're making the calculation you think you are. If all else fails, just use **count( )** and write out the whole expression.

With Fathom's tests and estimates, you will be able to do many of the exercises in the rest of the chapter. Although they will make some things easy, be sure to do at least *one* exercise using simulation; after you have done so, use the orthodox tests to check it. Remember, also, that if your situation doesn't meet some criteria for using a t or z procedure, the logic of the simulation may still hold.

Exercises

13.25 LYME DISEASE. You will need to enter summary statistics into an interval estimate. Use **Difference of Proportions**. You will have to come up with good names for the **Attributes** and **Categories**.

13.26 FEAR OF CRIME AMONG OLDER BLACKS. Same as 13.25.

13.27 POLICE RADAR AND SPEEDING. Note that part (a) is a one-sample question!

13.28 INFORMATION ONLINE. Not everything here is two-sample. Be sure to use the right tool for each sub-question!

13.34 REDUCING NONRESPONSE? Be careful that you know exactly what you're testing when you enter numbers into the computer!

13.38 STEROIDS IN HIGH SCHOOL. As always, be sure to label your categories and attributes well; it will help you keep things straight. Also, be sure to use the right tools.

13.44 EACH DAY I AM GETTING BETTER AT MATH. Use **EX13_44 Better at math.ftm**. You will need at least one computed column.

14 Inference for Distributions of Categorical Variables: Chi-Square Procedures

In the last part of Chapter 13, we saw how to compare proportions. That works in a 2-by-2 situation, but what if there are more than two categories? A simple difference of proportions doesn't make sense; we need a new tool.

That's where the chi-square (χ^2) statistic comes in. It's an ingenious way of calculating how much a categorical distribution differs from a hypothesized distribution (which leads us to the goodness-of-fit test), or alternatively, how much a two-way (two-variable) distribution departs from independence.

14.1 Test for Goodness of Fit

We begin with the test for goodness of fit. In this test, we compare a single-variable categorical distribution to some hypothesized distribution. The textbook begins with accident data and days of the week. The (null) hypothesized distribution is that the accidents should be roughly uniform over the seven days, so each would have a probability of 1⁄7. (Note: The hypothesized probabilities don't have to be equal as they are here.)

SUSPENSION OF DISBELIEF FOR INSTRUCTIONAL PURPOSES. Don't be misled by the context in the textbook! The opening accident data have nothing directly to do with cell phone use, but they do give us data on which we can use this test. You might also rightly complain that you don't need statistics to tell that there are fewer accidents on weekends—it jumps right out of the table. True. But knowing the answer means we can check up on our reasoning as we develop the tool.

Example 14.1 Auto accidents and cell phones

This time, let's begin with the Fathom test and do the simulation later.

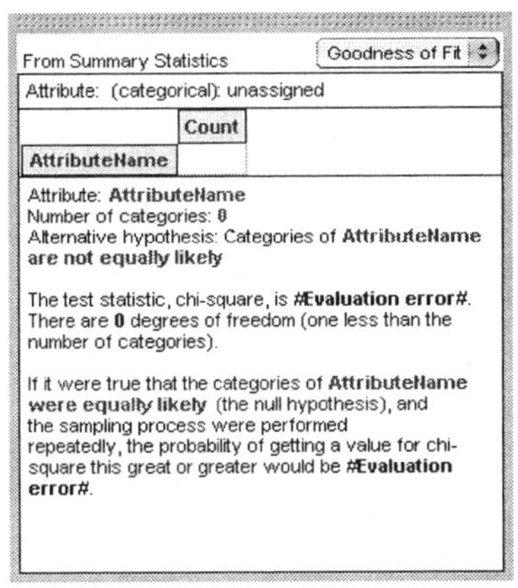

- ❖ In a new Fathom document, drag a test off the shelf and choose **Goodness of Fit**. It will look like the illustration at right.
- ❖ We're going to edit the blue text. At the top, change **Attribute** to **day** and set the number of categories to 7. The table will change dramatically to give you a table where you can enter the names of the days and the data. Do so. Your test object will look like the next illustration:

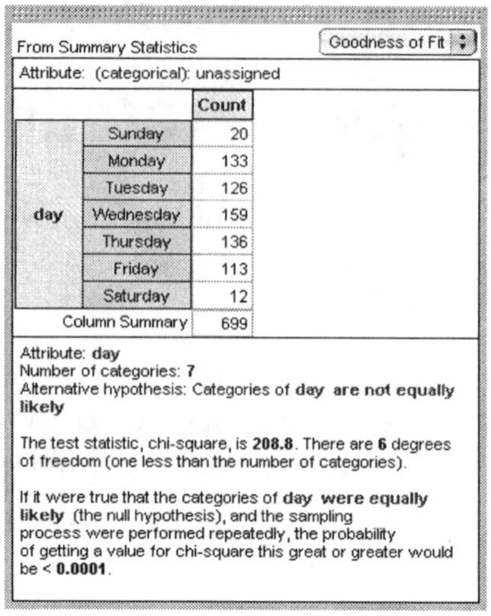

...and you're done! The *P*-value is really small, as we expected.

Looking at Mississippi data

In the example, you entered summary data. Now let's look at raw data from the National Fatality Analysis Reporting System (FARS).[1] The file you will use contains one record for each fatal auto accident that occurred in Mississippi in 2004.

❖ Open the file **14 Mississippi Crashes.ftm**. Drag a new test off the shelf and choose **Goodness of Fit**.

❖ Drag the day-of-week text variable, **DOW**, from the table to the slot at the top of the test, where it says **Attribute (categorical): unassigned**. If you also make a graph of that variable, you'll see this:

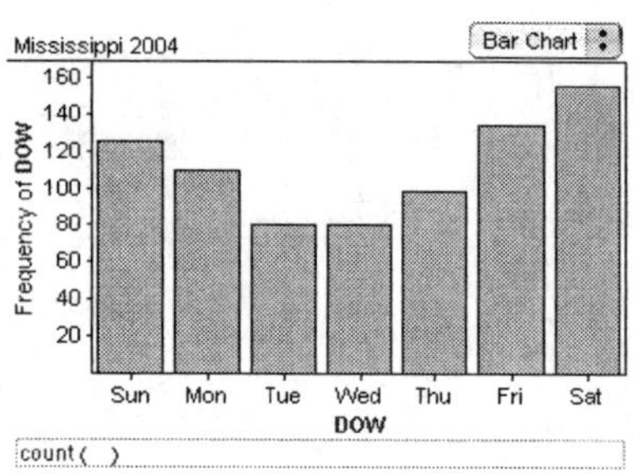

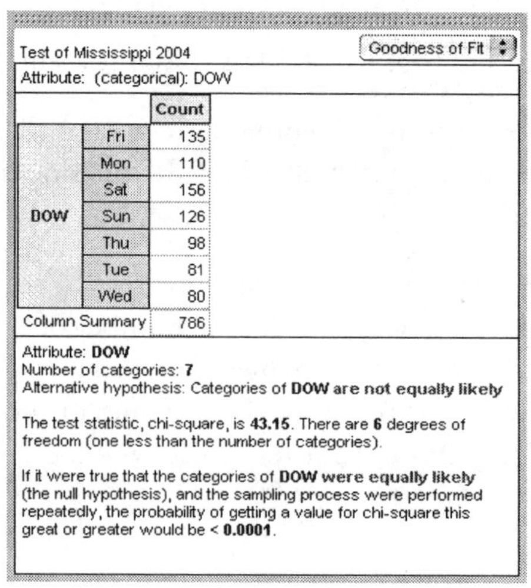

1 We got the data from **http://www-fars.nhtsa.dot.gov/main.cfm**.

This time we didn't enter the data directly into the table because we had a separate case for each crash.

Although the days are more "even" than in the textbook example, the test still confirms (with $P < 0.0001$) what our eyes tell us: We have strong evidence against the null hypothesis that crashes are equally likely on each day of the week.

As an alternative, drag the **month** number to the "slot" at the top of the test and graph it as well. The data are consistent (at $P = 0.13$) with fatal accidents in Mississippi being distributed uniformly among the months.

TOOLBOX CHECK. These data are a census, that is, they constitute all of the fatal accidents in Mississippi in 2004. So what is our population? If we say "all accidents in Mississippi," we're taking the ones in 2004 as an SRS of all years. If we say, "all U.S. accidents in 2004," we're taking the Mississippi accidents as an SRS of those from the whole country. Neither of these is reasonable. Nevertheless, researchers use tests like these; in effect, the population is "All sets of accidents that are uniformly distributed across days (or months)," and we are showing that the set of Mississippi data from 2004 is *not* an SRS of such a data set.

Simulating to find the chi-square distribution

What does the chi-square distribution look like? Fathom has functions parallel to the ones we have used for the Normal and t distributions:

chiSquareDensity(x, df)	The density of the χ^2 probability density function at value x, with degrees of freedom df.
chiSquareCumulative(x, df)	The cumulative density up to x.
chiSquareQuantile(c, df)	The value of χ^2 below which the proportion c of the distribution lies. Use this to find critical values, for example, with $c = 0.95$.

But instead of simply plotting these functions, let's generate a sampling distribution and *then* plot the density function to see how it matches. We'll simulate a truly uniform distribution of days and see what chi-squared values we get. Bear with this; there is a surprise in store.

- In a new Fathom document, make a new table with one attribute, **day**.
- Give **day** the formula **randomPick("Sun","Mon","Tue","Wed","Thu","Fri","Sat")**. Be sure to put quotes around all of the text strings and commas between them.
- Give the collection 100 cases (i.e., 100 rows). One hundred random **day**s appear. Graph the data to make sure they look reasonable. (The days won't be in order. Doesn't matter.)
- Open the collection inspector and go to the **Measures** panel. We'll make a lot of measures. The first eight will make writing the ninth formula easier.

❖ Use the letter **O**, not the number **0**, and make **O1** equal to **count(day = "Sun")**, make **O2** equal to **count(day = "Mon")**, and so forth up to **O7**. These are the **O**bserved values. Then make **E** (i.e., **E**xpected) equal to **count( )/7**. Your document will look like this:

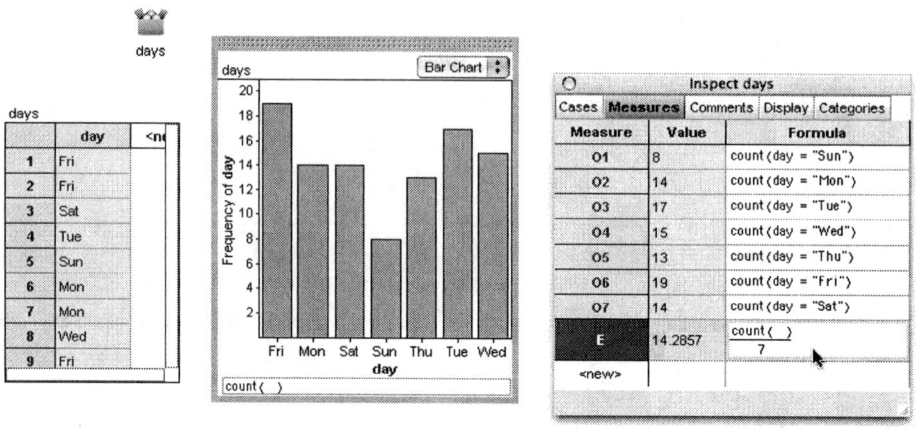

❖ Now for the last one, **chisq**. Make the measure and open the editor. Its formula is

$$\chi^2 = \sum \frac{(O-E)^2}{E}$$

but we have to write out the summation. So enter **(O1 – E)^2/E + (O2 – E)^2/E +** and so forth, until you're done. It's not as bad as it looks, especially if you copy and paste. Just make sure you change the digits after the **O**s!

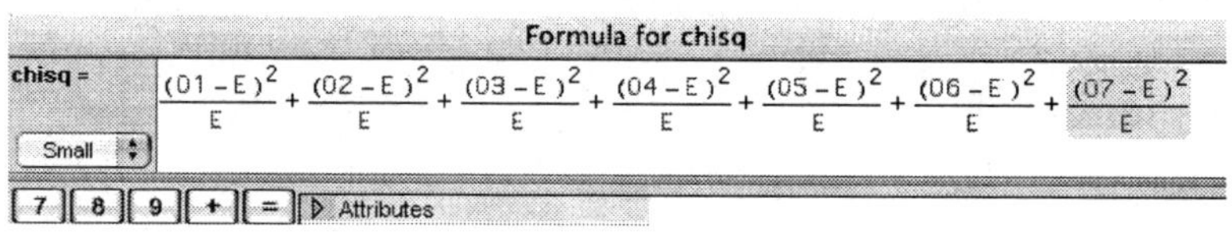

Make sure your value for χ^2, namely **chisq**, looks right.

❖ Collect at least 500 measures and plot the resulting distribution of **chisq**. We collected 500 and got this histogram:

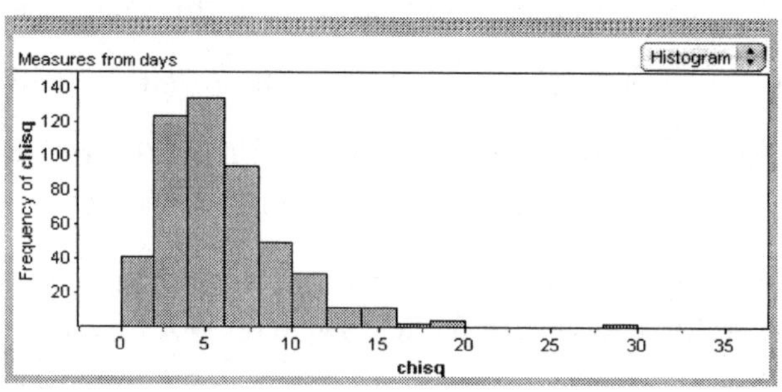

❖ Does it match the theoretical distribution? Set the histogram to have a density scale (context>**Scale>Density**) and then choose **Plot Function** from the context menu.

- Enter **chiSquareDensity(x, 6)** because with 7 days, there are 6 degrees of freedom. The function should match pretty well.
- Do the degrees of freedom really matter? Make a slider and call it **df**.
- Edit the function formula to be **chiSquareDensity(x, df)**.
- Play with the slider to see how the function changes with different degrees of freedom. You should see that it matches your distribution best around **df** = 6.

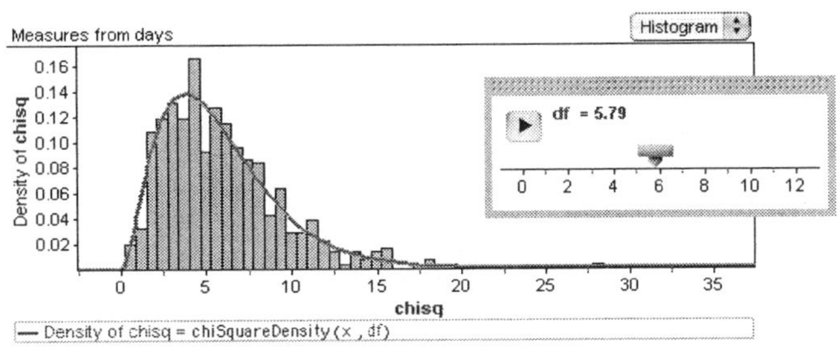

This looks nice and smooth, but the histogram hides a dirty secret.

- Change the graph back into a dot plot.
- Zoom into a spot on the axis near **chisq = 5**, either by dragging or using the zoom tool (in Windows, hold down CONTROL; on the Mac, use OPTION). Zoom way in until you can see that the dots line up in little piles.

In a given situation, *not all values of* χ^2 *are possible*. This is not like the times when we simulated by choosing a random number from a Normal distribution. Those numbers could be anything. And as long as there are lots of cases and few categories (few "bins"), the χ^2 distribution is pretty smooth. But when the number of cases goes down and the number of categories goes up, the graininess of the distribution can affect the *P*-values you get. This is one of the reasons you need 80% of the expected counts to be at least 5.

Unequal expected counts

So far we have studied only equally likely situations. Let's look at an exercise from the textbook that breaks that mold:

14.3 COURSE GRADES. In this problem, the population of grades is given with various probabilities: *P*(A) = 0.32, *P*(B) = 0.41, *P*(C) = 0.20, and *P*(D/F) = 0.07. We want to see if the professor's grades could plausibly come from that discrete distribution. How do you deal with this in Fathom?

- In a new document, make a test and choose **Goodness of Fit**.
- In the test, change **AttributeName** to **grade** and set the number of categories to **4**. Enter the names of the categories (e.g., **A**) and their counts (**22**). At this point, your test will look like this:

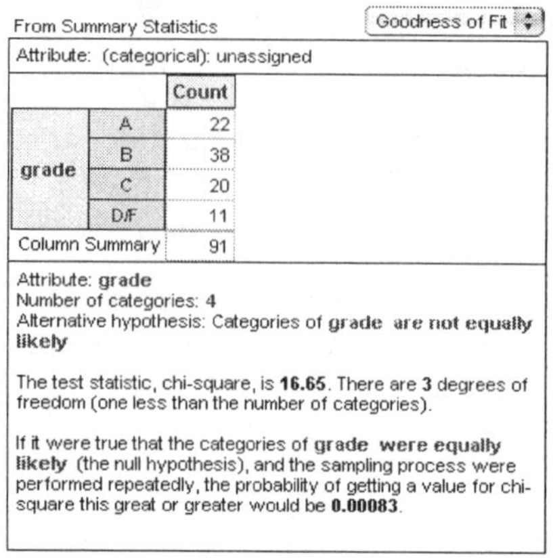

We have a problem: The null hypothesis here is that the grades were *equally* likely. That's not right; our null hypothesis is that the grades follow the distribution of the rest of the class.

- ❖ Click on the blue text that says **were equally likely** and choose **had probabilities given above** from the menu that appears.
- ❖ A column appears showing equal probabilities. Edit them to match the distribution we're after. You can also use the context menu to show the distribution of χ^2 in this situation. When you're done, the test and graph should look like this:

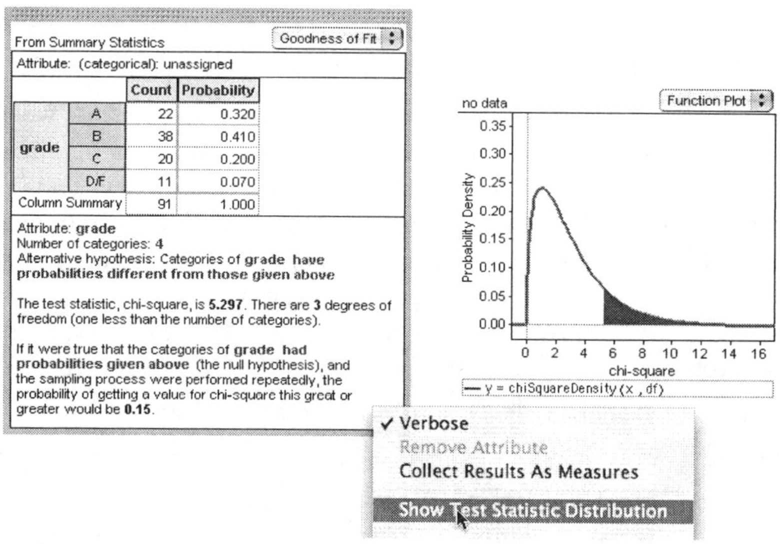

These tools can help you with most of the exercises in this section. Some notes:

14.9 Is your random number generator working? In this exercise, use the Fathom formula **randomInteger(9)** to get random numbers from 0 to 9. You can graph these any way you like. To get a table of how many of each digit you get, hold down SHIFT as you drop your variable name onto a summary table.

At this point, you can enter the counts (and the expected values) into a new table and proceed with the exercise, computing χ^2, and so forth. (You may need an additional summary table to compute the sums of the terms.)

Alternatively, you could context-click on the summary table and choose **Create Collection from Cells**. The resulting collection will treat each summary cell as a separate case. If you add another formula to the table to compute the expected values, you can continue with the exercise without re-entering the counts.

Or you can use a test object off the shelf, and let Fathom do the test.

14.2 Inference for Two-Way Tables

We have been looking at categorical one-variable distributions and comparing them to hypothesized values. In this section, we'll look at two categorical variables instead of one. We often display these data in a table where the categories for one variable are rows and the other variable sits in the columns. Such a table is a *two-way table*, and if it has *r* rows and *c* columns, we call it an *r by c* table.

To study such a situation, good old χ^2 is a good statistic to use—you're comparing an existing distribution to a hypothesized one, and

$$\chi^2 = \sum \frac{(O-E)^2}{E}$$

is a good measure of how far "off" the observed data are from expectations. In the two-way table, however, instead of saying that all categories are equally likely or that they should follow some *a priori* set of probabilities, we will compute the expected values from the data, assuming that knowing the value of one variable has no effect on how well we know the other. Put another way, we assume there is no *association* between the variables.

How you compute these expected values is ingenious: You take the product of the marginal counts and divide by the grand total; or, alternatively, you take the product of the marginal *probabilities* and multiply by the grand total. If you do this, the pattern of proportions is the same in each column; likewise in each row.

Example 14.4 Does background music influence wine purchase?

Let's look closely at this example from the textbook, beginning with some exploratory data analysis.

- ❖ Open the file **EG14_04 music and wine.ftm**. We have created a collection with 243 cases, one for each bottle of wine sold. The variables are **wine** and **music**.
- ❖ Make a new summary table; drag **music** (the explanatory variable) to the horizontal spot at the top to make columns. Drag **wine** (the response variable) to the vertical spot to make rows. The result should be a table of counts like the one in the textbook (though the columns may be in a different order).

wine study

		music			Row Summary
		French	Italian	None	
wine	French	39	30	30	99
	Italian	1	19	11	31
	Other	35	35	43	113
Column Summary		75	84	84	243

S1 = count( )

❖ Make a new graph. Again drag **music** to the horizontal axis and **wine** to the vertical. You'll see a fairly confusing split bar chart. Let's get rid of the wine. Right-click on the graph and choose **Remove Y attribute: wine**.

❖ Now drag **wine** back to the graph. This time, take it to the horizontal axis and *drop it on the plus sign that appears*, so it gets added to the graph instead of replacing **music**. The illustration shows the drop site and the resulting graph.

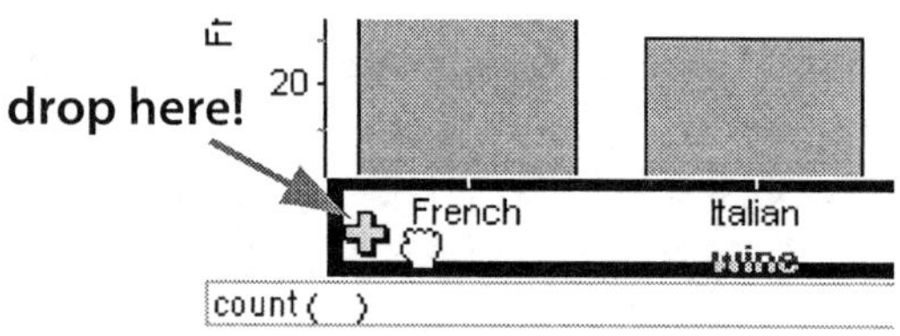

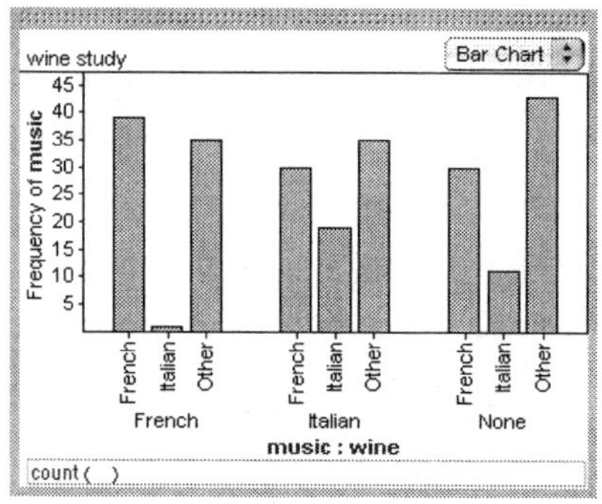

For many people, it's easier when looking at this graph to see how the patterns are different from one another. It's easy to see, for example, how French music seems to suppress Italian wine.

❖ Use **Undo** (CONTROL-**Z** or **⌘-Z**) to get rid of **wine** again. This time, drag it to the graph and drop it in the *middle* of the graph, on top of the bars. Now you get a stacked bar chart. Change it to a ribbon chart; both appear in the next illustration:

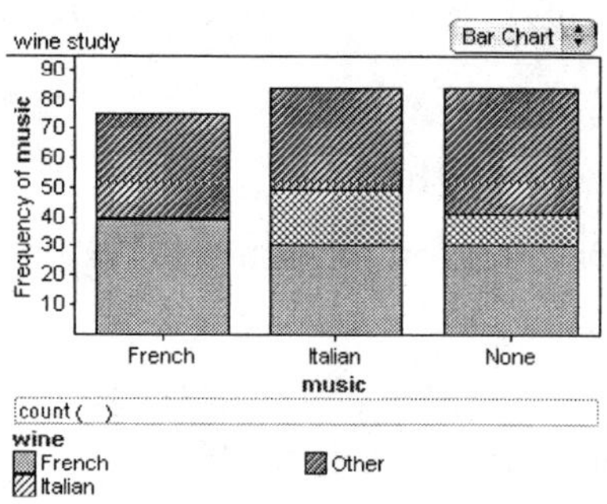

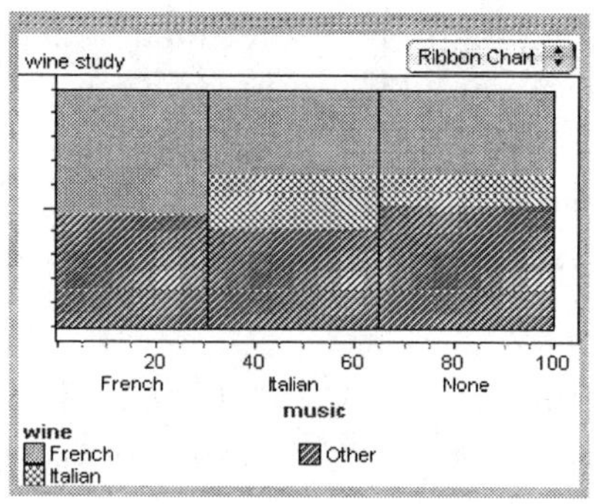

The Glories of Ribbon Charts, Revisited

The ribbon chart, though probably the least familiar to you, is worth getting to know. Each bar stretches the full height of the graph. The widths of the vertical bars are proportional to the column totals, which represent the total number of bottles sold during each kind of music. That way, the area of each segment is proportional to its frequency and you can easily see how the wine sales change with the music.

In the null situation, each column is divided in the same proportions. This is equivalent to saying that the columns come from the same population. And if the columns (wine sales) are identical no matter what music is playing, the two variables are *independent*; this is why the test for independence and the test for homogeneity are the same.

The "visual test for homogeneity" is the same as on page 180 in Chapter 13 in this guide: If the horizontal lines—the cracks between the segments of the bars—line up, the proportions are equal. If they don't line up, there's a gap. And the bigger the gap, the bigger the difference.

Example 14.7 Are music and wine linked? The chi-square test

We can see that the proportions are not the same in the different columns, but could that be just by chance? Let's look at how to perform the test in Fathom. Fathom will take care of all the calculations, including the expected values.

We continue with our wine-and-music data set, **EG14_04 music and wine.ftm.**

- ❖ Drag a new test off the shelf. Choose **Test for Independence** (because it's the same as a test for homogeneity).
- ❖ Drag **music** to the top slot in the test and **wine** to the bottom one. The order does not matter, really, but the first variable defines the columns, which we often associate with the explanatory variable. Fathom completes the test:

Test of wine study — Test for Independence

First attribute (categorical): music
Second attribute (categorical): wine

		music			Row Summary
		French	Italian	None	
wine	French	39 (30.6)	30 (34.2)	30 (34.2)	99
	Italian	1 (9.6)	19 (10.7)	11 (10.7)	31
	Other	35 (34.9)	35 (39.1)	43 (39.1)	113
Column Summary		75	84	84	243

First attribute: **music**
Number of categories: **3**
Second attribute: **wine**
Number of categories: **3**
Alternative hypothesis: There is an association between **music** and **wine**

The test statistic, chi-square, is **18.28**. There are **4** degrees of freedom (the number of rows minus one times the number of columns minus one).

If it were true that **music** were independent of **wine** (the null hypothesis), and the sampling process were performed repeatedly, the probability of getting a value for chi-square this great or greater would be **0.0011**.

The numbers in parentheses in the table are expected counts.

With $P = 0.0011$, we reject the null hypothesis. Notice that the test gives you a nice table and the expected cell counts. It's a good thing, too: Rejecting the null hypothesis in this kind of test only says that the distributions of the variables are not the same; to tell *how* they're different, you need to look at the details.

Exercises

You can use Fathom for many of the rest of the exercises in this chapter. We'll go over one carefully so you can see how to enter summary statistics into this kind of chi-square test.

14.11 HOW TO QUIT SMOKING, I. But once you have done this exercise without using a computer-generated test, you can check your work:

- ❖ In a new document, drag a test off the shelf and choose **Test for Independence**.
- ❖ Edit the blue text: change the **FirstAttribute** to **Result** and give it **2** categories. The **SecondAttribute** is **Treatment**, with **4** categories.

The categories in **Treatment** are **nicotine patch**, **drug**, and so forth. The categories for **Result** are **Success** and **Failure**. *Note that you will have to compute Failure numbers from the data in the table!*

- ❖ Expand the test to show the table,
- ❖ Enter category names by double-clicking them and editing.
- ❖ Enter the counts in the table. You can check your "Failure" arithmetic by looking at the marginal totals. That is, the total number of subjects for **Nicotine patch** must be 244.

The illustration shows what you might see. There is clearly an association (because the *P*-value is so small) but note that the chi-square test does not tell you which treatment is best. For that you need to look at the expected counts (in parentheses), at ratios, or at a good graph.

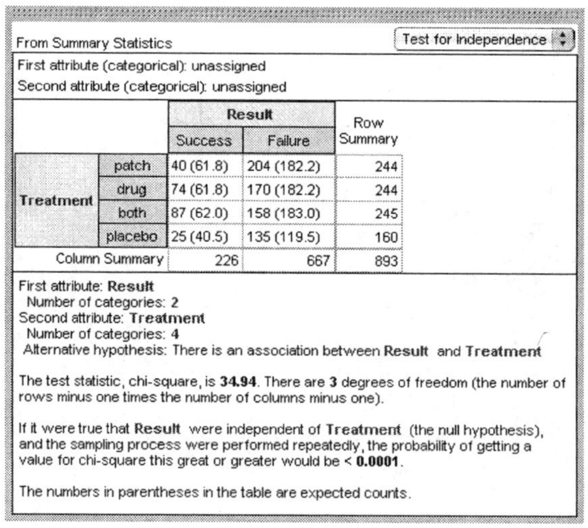

From Summary Statistics — Test for Independence

First attribute (categorical): unassigned
Second attribute (categorical): unassigned

		Result		Row Summary
		Success	Failure	
Treatment	patch	40 (61.8)	204 (182.2)	244
	drug	74 (61.8)	170 (182.2)	244
	both	87 (62.0)	158 (183.0)	245
	placebo	25 (40.5)	135 (119.5)	160
Column Summary		226	667	893

First attribute: **Result**
Number of categories: 2
Second attribute: **Treatment**
Number of categories: 4
Alternative hypothesis: There is an association between **Result** and **Treatment**

The test statistic, chi-square, is **34.94**. There are **3** degrees of freedom (the number of rows minus one times the number of columns minus one).

If it were true that **Result** were independent of **Treatment** (the null hypothesis), and the sampling process were performed repeatedly, the probability of getting a value for chi-square this great or greater would be < **0.0001**.

The numbers in parentheses in the table are expected counts.

15 Inference for Regression

Back in Chapter 3 you learned about linear regression and the least-squares line. To make things simpler, we assumed the slope and intercept of that line were certain. They're not. Just as the mean of a sample is not the mean of a population, the slope in a graph of a sample is not the "true" slope from the population: Sampling variability makes each sample a little different.

As a result, the slope[1] you get from a regression calculation has a confidence interval, and you can test whether that slope is significantly different from some special value (namely, zero).

Example 15.1 Crying and IQ

We'll begin with a little exploratory data analysis, then introduce you to the relevant Fathom tools.

❖ Open the file **EX15_01 Crying and IQ.ftm**. It contains the crying data from the textbook.

❖ Make a scatterplot using **Crying** as the explanatory (x) variable and **IQ** as the response.

❖ Put a least-squares line on the plot. You can see that there is a positive slope, but that there is quite a bit of variation about the line.

This much you have done before. You can read the slope and intercept and r^2 (about 0.2), which tells you about the strength of the relationship. Let's move on:

❖ Drag a new *model* off the shelf. This is the third (and final) inference "object" in Fathom, joining the *test* and *estimate*. Choose **Simple Regression** from the model's corner menu.

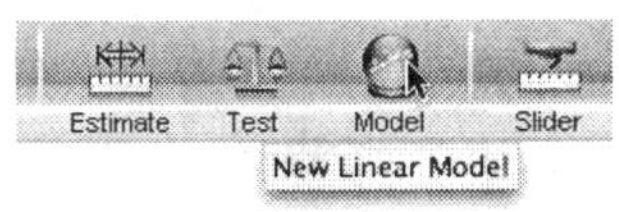

❖ Drag **IQ** to the top slot (the *response attribute*) and **Crying** to the second slot (the *predictor attribute*).[2] Now the graph and model look like this:

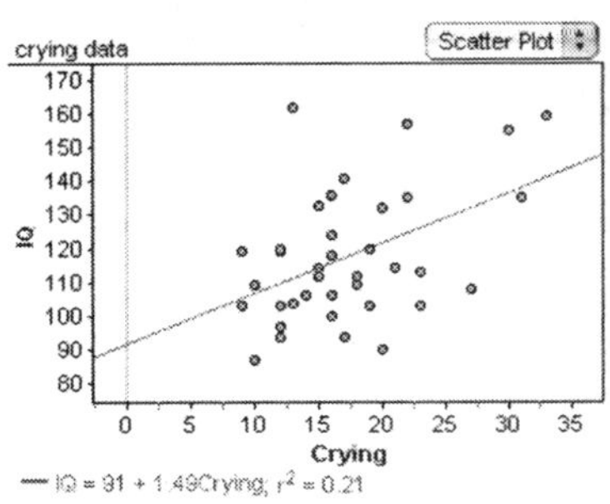

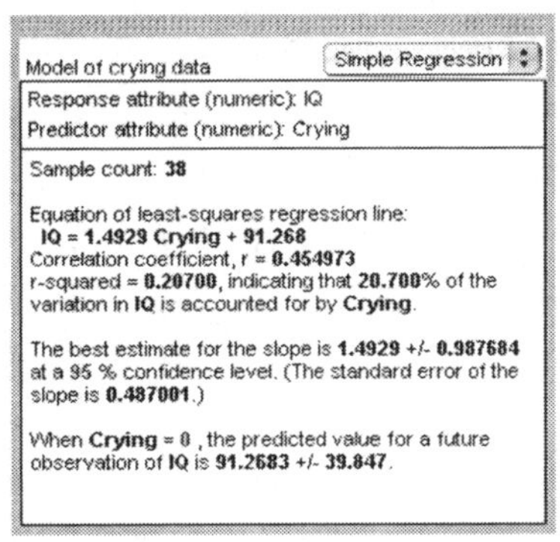

1 Or intercept, but we'll only do slope here.

2 Fathom, in addition to using *attribute* for *variable*, uses *predictor* instead of *explanatory*. So where you read "predictor attribute," you can mentally substitute "explanatory variable." Or maybe just "x."

Notice that the two formulas for the best-fit line are the same. The least-squares line *is* a simple regression. The model box also gives you more quantitative information:

- ❑ You get r instead of just r^2. Here it's about 0.45.
- ❑ You get an interval estimate for the slope; in this case it's about 1.5 ± 1.0. You can change the confidence level for that interval.
- ❑ You get the standard error of the slope (which gets discussed in the textbook).
- ❑ Given a value for a predictor, you can have Fathom compute a response. For example, edit the blue number—which is **0** by default—and make it **20**. Press TAB or ENTER, and you will see the range of IQs to expect from an infant who cries at level 20.

Exercises

15.1 AN EXTINCT BEAST, I. Enter the *Archaeopteryx* data by hand. Plot a graph, add a least-squares line, and make a Simple Regression model (i.e., use a model from the shelf). Then answer the question in the textbook.

15.2 IDEAL PROPORTIONS, I. Unlike other tests and estimates, *you cannot enter summary data and get Fathom to make a linear model.* Just use the material from the textbook or use some raw data in **EX15_02 Ideal proportions.ftm**.

Looking Closely at Residuals

15.3 SPARROWHAWK COLONIES. The file **EX15_03 Sparrowhawks.ftm** has the data. You can make the graph with an LSRL, and you can make a residual plot in that graph to assess the spread about the line. But what if you want to look at the residuals in detail, as in the textbook?

- ❖ Make a new column and give it a suitable name such as **resid**.
- ❖ Give it this formula: **linRegrResidual(PctReturn, NewAdults)**. The form of this function is *linRegrResidual(predictor, response)*. The values for the residuals appear in the new column. You can graph them like any other data:

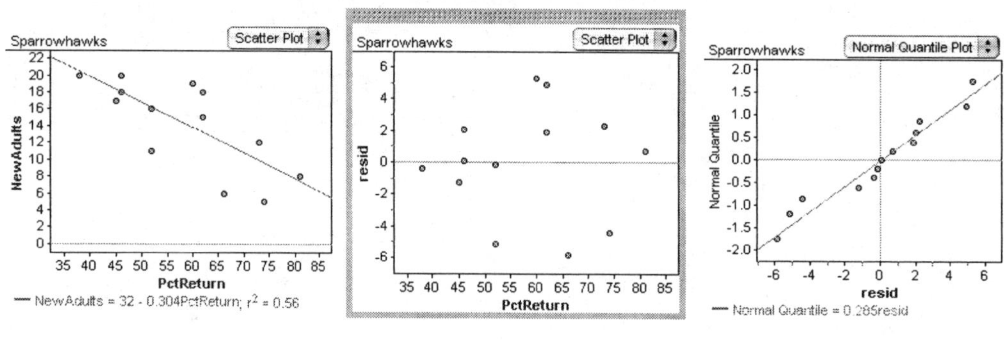

The illustration shows the original LSRL graph; a plot of **resid** against **PctReturn**, the predictor; and a Normal quantile plot of just the residuals to show that they're consistent with Normality (requirement (d)).

15.4 DOES FIDGETING KEEP YOU SLIM? Use the file **EX15_04 Fidgeting.ftm** and apply what you've just learned.

15.6 COMPETITIVE RUNNERS, I. Use the file **EX15_06 Runners.ftm**. For part (a), make a new column for residuals as described above. For part (b), a summary table works really well.

15.7 THE LEANING TOWER OF PISA. Use the data in **EX15_07 Leaning Tower.ftm**. Use Fathom graphs and a model to answer the questions in the textbook.

15.8 PREY ATTRACT PREDATORS. Use **EX15_08 Prey and Predators.ftm** for the data. Notice how we had to enter the data: one case per observation, instead of the way the table is laid out in the textbook.

Competitive runners digression

Aloysius looks at the graph of the data in **EX15_06 Runners.ftm**, which looks mighty linear, and says, "Well, *duh*. It has to be this way: If you take twice as many steps per second, you'll go twice as fast." But Nathalie takes him down a notch when she asks, "But then why is there an intercept?"

Why is Nathalie's comment relevant? How long (in feet) is each stride and how does that stride length depend on overall speed or on the number of steps per second? That is, when the runners go faster, do they take longer steps or shorter steps? Is that relationship linear? How can you tell?

After exploring all that, what is the *meaning* of the slope on the original graph and why does it have the units it does?

Testing the hypothesis of no linear relationship

Even if there is no linear relationship, your data will show some slope when you do a regression, just from random fluctuations. This suggests an idea for a null hypothesis—that the true population slope β is equal to zero. We can test a data set against that null hypothesis using Fathom's test object, choosing **Test Slope** from the corner menu. If we reject this null hypothesis, that means there is evidence that the slope is "real." As explained in the textbook, the machinery for this is really that of a t-test; we use the slope's standard error just as we would a standard error of the mean.

Example 15.6 Crying and IQ again

Let's learn the mechanics of how to test for slope:

- ❖ Open the file **EX15_01 Crying and IQ.ftm**.
- ❖ Drag a new test off the shelf and choose **Test Slope** from its corner menu.
- ❖ Drag **Crying** to the predictor (second) slot and **IQ** to the response (first) slot. The test fills in giving you the result that $P = 0.0041$. We can reject the zero-slope hypothesis with confidence:

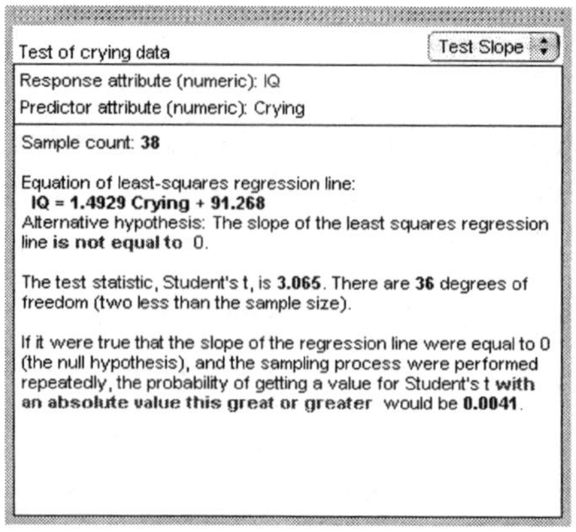

You can, as usual, change this 2-sided test to its one-sided counterpart by clicking on the blue text—which turns into a menu—and making an appropriate selection.

Note that you cannot change the slope you're testing *against*. You can only use β = 0 as a null hypothesis; no other slopes are allowed. That said, consider "ideal proportions" data like the set in **EX15_02 Ideal Proportions.ftm**. Suppose you wanted to test whether the slope of the **height-armspan** relationship was different from 1.00. There are at least two strategies:

1. Make a linear model and see if the confidence interval for the slope includes 1.00. (It doesn't at the 95% level).
2. Make a new variable (let's call it **excess**) and set it equal to **(height – armspan)**. Then test whether the slope on the **excess-armspan** graph is zero or not.

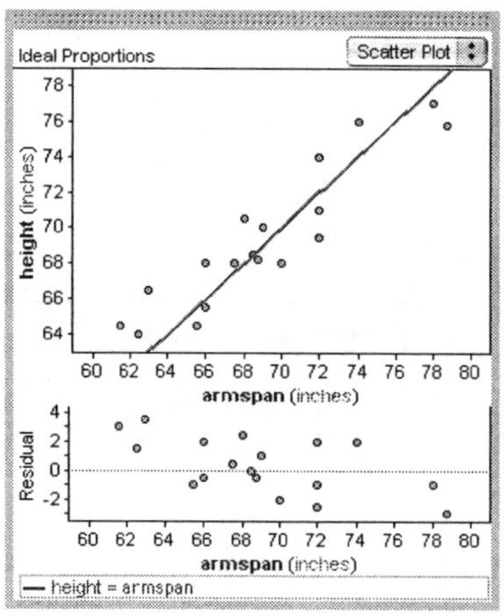

This illustration shows the data. We have plotted the line **height = armspan** and the residuals from that line. Notice how the residuals trend downward. Our new variable **excess** is the same as this residual.

You can use the file **EG15_07 Beer and BAC.ftm** to look at Example 15.7.

Exercises

15.11 An extinct beast, III. Open **EX15_01 Archaeopteryx.ftm** and use it to answer the questions. You will need to use both a model (to get a confidence interval) and a hypothesis test.

15.12 Competitive Runners, II. Use the file **EX15_06 Runners.ftm** and make a linear model set to **Simple Regression**. Use that model—and no graph—to answer parts (a) and (b).

15.13 Are jet skis dangerous? Use the file **EX15_13 Jet skis.ftm**.

15.14 Is wine good for your heart? Use **EX15_14 Wine consumption.ftm**.

15.15 Beavers and beetles. Use **EX15_15 Beavers and beetles.ftm**. Then use your Fathom tools to answer the questions. Compare the numbers to those in the Minitab printout in the textbook.

15.16 beaver and beetle residuals. Use the file from the previous exercise. Use the **linRegrResidual( )** function to get the residuals and then make another new column where you compute the standardized residuals. Then proceed with the questions in the textbook.

These tools—the **Simple Regression** model and the **Test for Slope**—will help you with many of the remaining exercises in this chapter.

Simulating a test for slope

Throughout the second half of this guide, as we have been studying inference, we have used simulation to understand what tests and estimates really mean. Let's do that again for the test for slope. Our null hypothesis H_0 is that the true slope β is equal to zero. We will pretend the null hypothesis is true, and use simulation to collect a sampling distribution of *slopes*. They will naturally be clustered around zero. Then we'll compare our test statistic (the actual slope) to this distribution to see if it's unusual.

You can use scrambling to simulate a true null hypothesis as you did when you studied the test for independence in Chapter 14. After all, if you have two variables x and y, and you scramble x, then there is no systematic relationship between the variables. The slope will therefore be zero. You can do this with the familiar crying-and-IQ data:

- Open the file **EG15_01 Crying and IQ.ftm**. Using **Crying** as the predictor (explanatory) variable, compute the slope using any technique you want. You should get about 1.49. That is your test statistic, the one you'll compare to the sampling distribution.

- Right-click on the crying data collection and choose **Scramble Attribute Values**. In the new collection, **Scrambled crying data**, make a table; note that **Crying** is scrambled.

- ❖ Open the scrambled collection's inspector[3] and make a new measure, **slope**. Use the formula **linRegrSlope(Crying, IQ)**. See the value for the slope? It's close to zero. Good.
- ❖ Make a graph of the scrambled data and put a least-squares line on it.
- ❖ Open the scrambled collection by dragging its right edge so you can see the button labeled **Scramble Attribute Values Again**. Press it and watch the scrambled plot and slope change. Your file should resemble the one in the illustration.

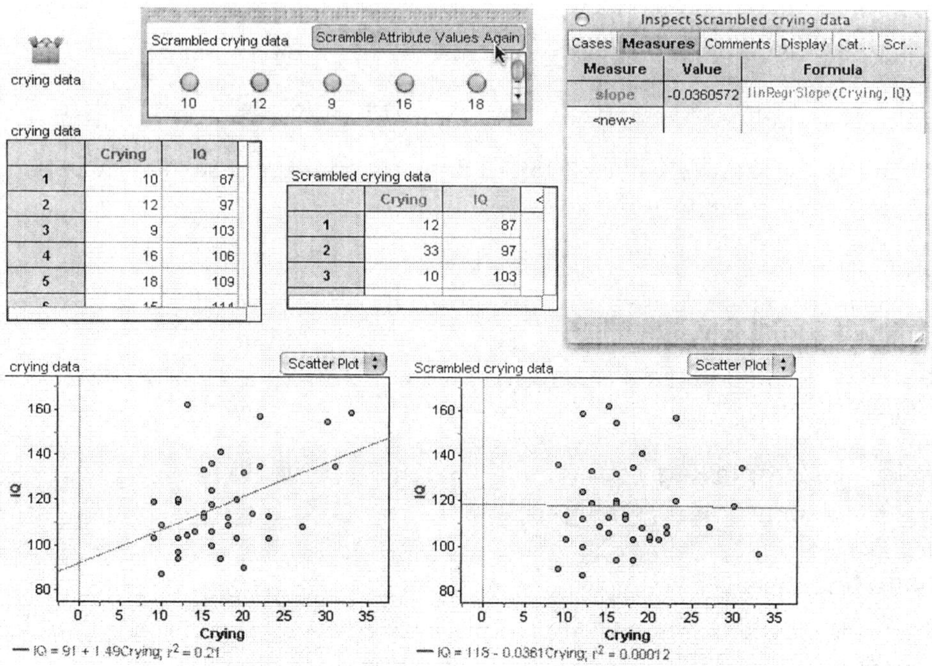

Pressing that button repeatedly, you should see that the slope from the real data (as in the graph at the lower left) is generally much larger than any you get from the scrambled data. They're the same numbers, of course, just mixed up. But that mixing, the enforcement of the null hypothesis, makes a big difference.

Now you can collect those slopes to make a sampling distribution.

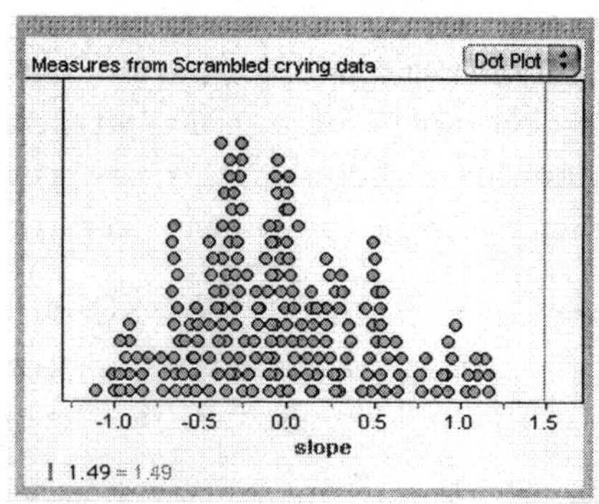

- ❖ Right-click on the scrambled collection and choose **Collect Measures**. Collect at least 100 and display the resulting distribution. If you wish, use **Plot Value** to show where 1.49 is, as in the illustration at right.

Note: if you have a reasonably fast computer, do this with animation *left on*. Watch how the distribution builds up, but especially watch the scrambled scatter plot and see how the LSRL wiggles around.

---

3 You could use the original collection's inspector instead, in which case its **slope** measure is yet another way to determine the slope in the data.

Special topic: Bootstrap interval for the slope

You can use the bootstrap technique to get an interval estimate for slope. You learned about it in Chapter 10; remember how it works? There are 38 cases in the collection, so sample 38 cases *with replacement*, compute the slope, and then collect those slopes as measures. You're not scrambling, so the data should still have an association, but it will vary a little depending on which cases get duplicated and which get left out. Look at the middle 95% to get an interval estimate for the slope—which you can compare to the confidence interval you'll get from Fathom's model object.

- Start with a fresh copy of **EG15_01 Crying and IQ.ftm**.
- Inspect the **crying data** collection and define the **slope** measure as before; use the function **linRegrSlope(Crying, IQ)**.
- Right-click on the **crying data** collection and choose **Sample Cases**.
- Inspect the new **Sample from crying data** collection and, using the **Sample** panel, set it to sample 38 cases with replacement. *Turn animation off.* Resample, and switch to the **Measures** panel where you can see the **slope**, which should vary about 1.49.
- Make a graph of the sampled data and put an LSRL on it, as before.
- Open the **Sample...** collection a little to see the **Sample More Cases** button; press it repeatedly to see the slope and the LSRL change.
- **Collect Measures** on the **Sample** collection; get at least 100 and make a graph of the distribution. (If you have the sample scatter plot and table open, they may flash and slow things down; you can get rid of them now—the graph and table, *not* the collection!)
- Use **Plot Value** to add lines to the new slope graph at the 2.5 and 97.5 percentiles. The first formula is **percentile(2.5, slope)**, though you can omit **slope** if you like, as in the illustration:

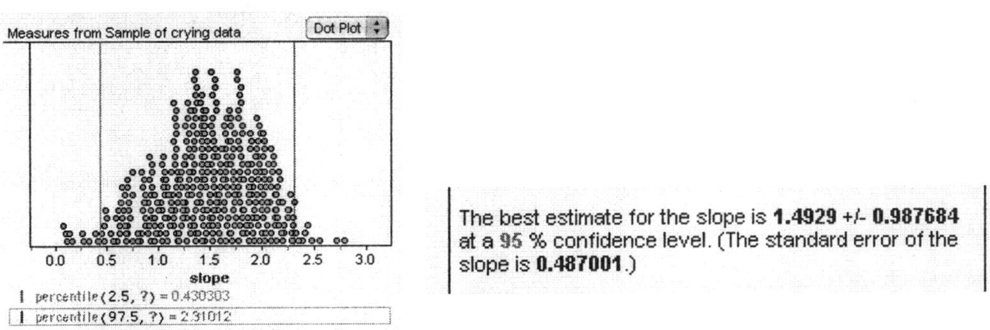

The bootstrap interval is [0.43, 2.31]. The illustration also shows part of a Fathom model calculation, where you can see a 95% confidence interval of 1.49 ± 0.99 or [0.50, 2.48]. The bootstrap distribution is skewed slightly left and the lower interval reflects this. But neither interval includes 0.0. This is consistent with the hypothesis test.

Special topic: Test of correlation

Does more crying in an infant set the stage for a high IQ when it reaches age three, or is it that more intelligent children cry a lot and we just can't measure IQ until age three? Which is the cause and which is the effect? Or does something else explain both? Sometimes we can't tell.

When you can't tell which is the explanatory variable and which is the response, it may be better to look at the correlation, which measures of the strength of a linear association *without assigning variables to either role.*

In the test object, the very last thing on the menu is **Test Correlation**. It works as you would expect, testing the relationship between two variables against the null hypothesis that the correlation is zero. Simply drag two variable names to the two slots and the test fills in:

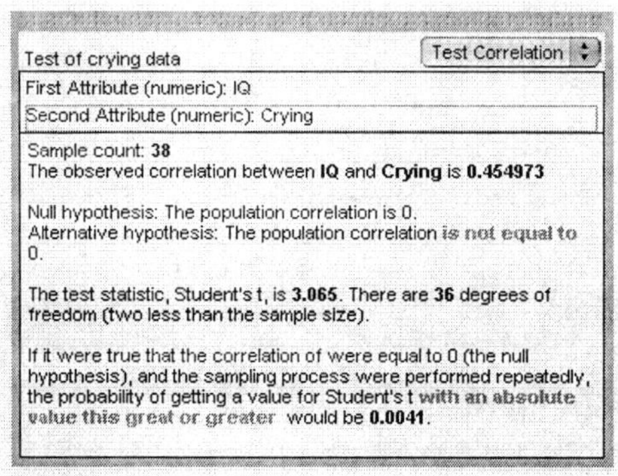

Unlike in **Test Slope**, the two variables are equivalent: You don't identify them as predictor and response. This is because the procedure for calculating correlation (unlike the procedure for regression) is *symmetrical*; you can switch the x and y variables and the formula does not change.

Despite that distinction, the *P*-value in the correlation test, 0.0041, is the same as that in the slope test. See if you can figure out if that is always true, and if so, why.